THE EVER-CHANGING MOON

Book One: First Footprints

Rob Bailey

For my dear wife, Scotty,
and our wonderful sons,
Robert, Alex, Nikolas, and Cameron

November 8, 1967. Apollo 4 waits on Pad 39A
the night before the first Saturn V launch. Photo credit: NASA.

Acknowledgments

My utmost gratitude goes to my wonderful wife, Scotty, who has endured my craziness since 1989, and to our sons Robert, Alex, Nikolas, and Cameron. They have graciously and patiently allowed me countless weekends and late-night hours to commit to this project since 2018.

I am honored to thank the many people who have made this possible. Five years ago, I never imagined I would contact Apollo veterans whom I have studied and admired all my life. I am constantly amazed at their willingness to help an unknown researcher. Yet they spent many hours answering numerous questions in person, on the phone, via Zoom, and through emails. They did not give mere stock answers but went into great detail. Without exception, the NASA veterans involved have been magnificent: Charlie and Dotty Duke, Gerry Griffin, Steve Bales, Doug Ward, Jerry Bostick, Tony England, David Reed, Bill Stoval, Jay Honeycutt, Tom Stafford, Jim McDivitt, Ed Fendell, Jim Lovell, Mike Dinn, John Saxon, Farouk El-Baz, Andy Foster, Wyckliffe Hoffler, Richard Stachurski, Gary Johnson, John Tribe, William Reeves, Chip Walraven, and Charles Clayton.

I would also like to express my gratitude to numerous talented historians who keep Apollo alive and have been very kind and helpful to me over the years: Colin Mackellar, Eric Jones, David Woods, Richard Jurek, Stephen Slater, J.L. Pickering, Kipp Teague, Andrew Chaikin, Danny Caes, John Stoll, Greg Wiseman, Jonathan Ward, Kathleen Phillips Esposito, Chris Calle, Katy Vine, Andrew Baird, Roger Launius, Alexandra Geitz, Jordan Whetstone, Scott Schneeweis, Robert Pearlman, Mark Graves, Paul Liberato, Mary Noel Black, and La-la Smith.

Many people have read drafts of chapters and offered excellent historical, stylistic, and editorial corrections: Bruce Schueneman, Colin Mackellar, Stefanie Bennett, Mark Pavlick, Linda Landreth, Gordon Landreth, Scotty Bailey, John Warren, Josh Feinberg, Collin Wood, Toby Herro, Tamara Yzaguirre, Kathleen Phillips Esposito, Leah Nass,

Josh Nass, Kylie Day, Charlie Duke, Dotty Duke, Steve Bales, Jerry Bostick, David Reed, Bill Stoval, Gerry Griffin, Doug Ward, Tony England, Jay Honeycutt, Farouk El-Baz, and Wycliffe Hoffler.

I am blessed to live in Corpus Christi, Texas, where several military flight instructors and engineers have added valuable insights: Josh Feinberg, Jeff Hale, Berch Abbott, Lloyd Ramey, Howard Irwin, and Collin Wood.

My goal has been historical accuracy. I take full responsibility for any mistakes. If you see anything needing correction, whether historical details or typographical errors, please contact me at rob@apollo16project.org. Thank you for your interest.

Neil Armstrong examines the footpad on a mockup
Lunar Module. Photo credit: NASA KSC-69P-578, Kipp Teague.

About
The Ever-Changing Moon

Imagine a world where people set aside political, religious, cultural, and philosophical differences to work together toward a common goal. That happened half a century ago when NASA went to the Moon. Four hundred thousand people around the world, with various gifts and personalities, mobilized for something bigger than themselves. During the historic first moonwalk, President Nixon told the astronauts, "For one priceless moment in the whole history of man, all the people on this Earth are truly one; one in their pride in what you have done, and one in our prayers that you will return safely to Earth."

Those memorable words reflected the emotions of many who watched and listened. Meanwhile, some scientists in Mission Control were annoyed Nixon was taking up valuable time for exploration. "Knock it off—we've got to get back to work!" Eventually, science would be a major factor in lunar missions, but that July evening in 1969 was a celebration of human accomplishment.

The Ever-Changing Moon is the story of extraordinary people involved with Apollo 16. *Book One: First Footprints* covers their contributions to human spaceflight through the monumental Apollo 11 mission. *Book Two: Man's First Journey to the Lunar Highlands* begins with the Apollo 13 training cycle, where geology gained more traction. It concludes with the treasures of Apollo 16, often regarded as one of the two lunar expeditions richest in discoveries.

The title is purposefully ironic. The Moon does not change, but our understanding of it does. The people and mission of Apollo 16 made significant contributions to space exploration and scientific knowledge. They disproved leading hypotheses about the Moon, forced new conclusions, and enjoyed the journey—I hope you do too.

Apollo 16 crew from left to right:
John Young, Ken Mattingly, Charlie Duke.
Photo credit: Houston Post.

Technical Details

First, thank you very much for reading. You will see many quotes and transcriptions of public air-to-ground conversations between astronauts and Mission Control in Houston. There are also transcripts of private conversations among flight crew members inside the spacecraft, captured by onboard recorders and not broadcast publicly. The quotes and transcripts on these pages come from mission audio and are my interpretation of those conversations. I have made slight alterations for clarity and out of respect for copyrighted material, but the speaker's intention is unchanged. Any errors in identifying the speaker or words spoken are my own. When audio was unavailable, I based quotes on NASA's onboard Command Module and Lunar Module transcriptions.

After flights, NASA released initial versions of mission transcripts which are found on the Johnson Space Center History Portal. Years later, Eric Jones met with moonwalkers, reviewed their lunar surface activities, supplemented corrected transcripts with commentary, and added supporting documents to produce the excellent Apollo Lunar Surface Journal. David Woods did a similar service for the flight portions of the missions, resulting in the Apollo Flight Journal. Both are tremendous works and highly recommended for anyone seeking more details about Apollo missions. Their corrected transcripts and commentary are copyrighted and hosted on the official history.nasa.gov website.

NASA Public Affairs Officers (PAOs) provided real-time mission commentary for the media, giving helpful summaries of flight activities. They serve the same purpose in this book and are indicated by italics. I have identified the PAO commentator by name in the transcripts. You can visit apollo16project.org to find more information about these men and samples of their memorable voices to help you identify them in other audio clips. I greatly appreciate Doug Ward spending numerous hours helping with the endeavor.

For more information about pictures, Arizona State University's website contains the Apollo Image Archive with high-resolution versions of all Apollo mission photographs, including those from the Panoramic and Mapping cameras. The ASU March to the Moon pages contain every image from the Mercury and Gemini missions. Kipp Teague's Apollo Image Gallery adds pre-mission and post-mission pictures, and you can find all of the mission photos in his Project Apollo Archive on Flickr and Facebook. J.L. Pickering at Retro Space Images is an excellent source for rare photographs, and his books are highly recommended. The Apollo Image Atlas on the Lunar and Planetary Institute website contains all Apollo images plus a useful search tool for feature names and coordinates.

Regarding video, Mark Gray's Spacecraft Films show complete in-flight TV broadcasts, onboard 16mm film, multiple angles of launches, full coverage of the moonwalks, plus other jewels.

For mission audio, John Stoll at Johnson Space Center has digitized and uploaded countless hours of priceless audio to archive.org. Apollo16project.org has an index correlating the Apollo 16 audio files with mission activities so you can listen to specific events. Ben Feist's Apollo in Real Time website is another fantastic labor of love, coordinating thousands of hours of audio from the flight controllers and astronauts for three missions, with more to come. Feist was a pivotal contributor to the award-winning Apollo 11 Documentary (2019, directed by Todd Douglas Miller), whose archival producer Stephen Slater has provided material to me I would never have seen.

My good friend Colin Mackellar is a wonderful source for hard-to-find audio, video, photographs, and technical information about the manned space program. His honeysucklecreek.net tribute to the historic tracking station is superb.

A glossary of terms is included in Appendix A to explain the technical jargon which characterizes life in Mission Control.

NASA divides many of its documents into sections and pages. For example, if a footnote references "Apollo 16 Mission Report, 14-17," it refers to section 14, page 17.

Regarding units of measure, spaceflight in the 1960s and 1970s referred to velocity in terms of feet per second and distance in nautical miles. On the lunar surface, NASA often employed the metric system, although some measurements used feet and inches. This book echoes the same conventions used by NASA for particular aspects of the missions. "Miles" refers to nautical miles unless otherwise noted. The book also uses terminology of the era such as "manned spaceflight."

My purpose is not to present a balanced view of the Apollo program; many excellent accounts have already been written for that purpose. My goals are to highlight the prominent figures of Apollo 16, introduce unfamiliar stories to the reader, and add to the research of this dynamic period of history. The amount of attention people or events receive may be disproportionate to their relative significance. For example, I mention the tracking station in Corpus Christi, Texas, several times. That does not mean it is more important than other stations. However, I have access to its local archives and want to add that information to the storehouse of Apollo history.

I live near Naval Air Station Corpus Christi and have been honored to attend many ceremonies when flight students are presented with their Wings of Gold and become Naval Aviators. The Commodore usually gives the wingees two charges: "Have fun, and pursue excellence." This book is the story of how a group of remarkable people did both with historic results.

Sea of Tranquility. Photo credit: NASA AS11-40-5877.

Lunar day 19, sketch by Erika Rix.
Courtesy of Erika Rix.

Contents

November 21, 1963. President John F. Kennedy, Mrs. Kennedy,
and Vice President Johnson at Brooks Air Force Base, San Antonio, Texas.
Photo credits: San Antonio Express-News (above)
Art Rickery, Time & Life Pictures (below).

Prelude

THE VOYAGE MAN ALWAYS HAS DREAMED ABOUT

APOLLO 11 LAUNCH: WEDNESDAY, JULY 16, 1969

CBS television cameras zoomed in on the mighty Saturn V (pronounced "Saturn Five") rocket as legendary anchorman Walter Cronkite began the live coverage:

> Man embarks today on history's greatest adventure. Here at the Kennedy Space Center in Florida, a massive vehicle waits to launch three American explorers into space, into the future, toward the lunar surface. The dawn of this day heralded the dawning of a new age. For with the first steps on the Moon, man's strides across the universe really begin. It's a time of exhilaration, reflection, hope, fulfillment, as a centuries-old dream starts toward reality.[1]

---

[1] Zellco321, "Apollo 11 Launch CBS News Coverage," video, 5:00:22, December 25, 2016, https://www.youtube.com/watch?v=ykRmdXzO9EI&t=5552s.

An enormous crowd gathered to witness the launch of the most powerful rocket ever built. Any Saturn V thundering off Pad 39A at Cape Kennedy was a spectacle, but there was something magical about this occasion—man's first attempt to land on the Moon.

Aboard were three courageous Americans: Neil Armstrong, Mike Collins, and Edwin "Buzz" Aldrin. One million spectators assembled at the Cape to experience the event in person. Though buffered from the launch pad by three miles, the ground shook beneath their feet as 120 decibels pounded them to the core.[2]

Vice President Spiro Agnew stood beside former President Lyndon Baines Johnson in the VIP section of the grandstands. In a nearby private viewing area, a group of astronauts watched with special guest Charles Lindbergh whose 1927 solo nonstop flight from New York to Paris was a turning point in aviation history.

One of those astronauts at the Cape was a thirty-three-year-old former Air Force test pilot whose voice would be forever linked to the dramatic lunar landing four days later. Charlie Duke was more than a spectator. He was heavily involved in the mission, representing the crew in crucial meetings during the hectic months leading up to the launch. Duke joined them a few days earlier, reviewing final details of the descent. In just six hours, he would assume the role of Capsule Communicator (CAPCOM), talking to Armstrong, Collins, and Aldrin from Houston's Mission Control.

Countless millions were glued to their television sets and radios, listening to the iconic voice of Jack King from the Cape's Apollo Saturn Launch Control:

T-minus fifteen seconds. Guidance is internal. Twelve.
Eleven. Ten. Nine. Ignition sequence starts. Six. Five. Four.
Three. Two. One. Zero. All engine running. LIFTOFF, WE
HAVE A LIFTOFF, thirty-two minutes past the hour. Liftoff
on Apollo 11!

---

[2] Mack R. Herring, *Way Station to Space: A History of the John C. Stennis Space Center*, (Washington, DC: National Aeronautics and Space Administration, 1997), 125. Mississippi Test Facility instruments measured 211 decibels at the base of its test stand when the first stage was fired. Spectators at Cape Kennedy were kept three miles away where the noise level decreased to an estimated 120 decibels. Two hundred decibels will kill a human. One hundred twenty decibels are roughly the equivalent of standing near a jet taking off, a chainsaw, a thunderclap, or an ambulance siren.

Shock waves blasted spectators and media several seconds later. An exuberant Cronkite said, "We're getting that buffeting that we've become used to…. What a moment! Man on the way to the Moon!"

Eyewitnesses shielded their eyes from the Sun as the mighty rocket soared out of view. Television cameras aided by telephoto lenses and chase planes tracked Apollo 11 for over five minutes until it was 270 miles downrange at an altitude of 82 miles. Cameras then focused on Vice President Agnew as he was escorted to the media by astronaut Tom Stafford. Meanwhile, Charlie Duke turned his attention to his first shift in Mission Control.

After saying goodbye to his wife, Dotty, Charlie hurried off to Patrick Air Force Base in his rental car. The normally-quick trip took about an hour due to traffic. He exchanged his civilian clothes for his flight suit and climbed aboard the white T-38 Talon, which the ground crew had already prepared for takeoff.[3]

The T-38 is a nimble airplane used by the Air Force for training pilots to fly jets. It has two seats and two engines and was the first supersonic trainer. Tom Stafford had been one of the plan's primary test pilots and recognized its potential for keeping an astronaut's flying skills sharp. He successfully lobbied for NASA to purchase a fleet of them.[4]

Duke was flying alone so he tossed his bag into the rear seat. The distance from Patrick AFB to Ellington Field in Houston was just inside the T-38 maximum range. Mileage was not a concern when flying east from Houston to the Cape since the prevailing wind was helping, but flying westward into the wind often required a fuel stop at Brookley Air Force Base south of Mobile, Alabama. The breeze was not strong on this historic day and Duke returned to Houston nonstop. To save fuel, he did not use the afterburner, opting instead for a "Military Power" takeoff. He advanced the throttle with his left hand to the detent, or stop, between the full standard power and the afterburner range. Soon the sleek jet was airborne.[5]

---

[3] Charlie Duke, interview by author, July 3, 2020.
[4] Space Center Houston, "Thought Leaders Series—Apollo 10 with Astronaut Tom Stafford," Houston, Texas, April 25, 2019.
Over fifty years later, astronauts still use the T-38 to maintain flight proficiency as they fly across the country for training and consulting with various spacecraft contractors.
[5] Charlie Duke, interview by author, July 3, 2020.

While Charlie Duke headed west, his friend and fellow astronaut Mike Collins was on his way to the Moon. Three hours after liftoff, Collins considered the commotion in their wake. "I'll bet the launch-day crowd down at the Cape is still bumper-to-bumper, straggling back to the motels and bars."[6]

Duke made it to the Mission Operations Control Room (MOCR) on the NASA campus with time to spare. Right on schedule, he took his seat at the CAPCOM console as Gene Kranz and his team of flight controllers began their shift in Mission Control. Duke pressed the button on his transmitter, enabling him to speak to the crew:

> **DUKE:** Hello, Apollo 11. Houston. Be advised your friendly White Team has come on for its first shift. If we can be of service, don't hesitate to call.
> **COLLINS:** Thank you very much.

The world watched and listened as Apollo 11 continued its audacious and inspiring mission, which Walter Cronkite eloquently described as "the voyage man always has dreamed about." He noted, "Neil Armstrong will take that first step, in more ways than one, and many things will never be the same again." [7]

Apollo 11 on the way to the Moon.
Photo credit: NASA KSC-69PC-397, Kipp Teague.

---

[6] Michael Collins, *Carrying the Fire: An Astronaut's Journeys*, (New York: Farrar, Strauss, and Giroux, 1974), 372.
[7] Zellco321, "Apollo 11 Launch CBS News Coverage."

1

THINGS WERE
DIFFERENT THEN

THE SPACE RACE

The Space Age began on October 4, 1957, when the Soviet Union launched the world's first artificial satellite, Sputnik 1. Though it weighed 184 pounds and was less than two feet in diameter, Sputnik could be seen through binoculars against a dark background just before sunrise and after sunset. Its menacing beeps sent a clear message to radio operators worldwide.

One such listener was a student at West Fargo High School in North Dakota named Tony England. He built his amateur radio by hand, combining craftsmanship with his love for electronics.[1] Ten years later, he became NASA's youngest astronaut, but on this day, England was an amazed tenth grader listening to the satellite beeping at him as it flew overhead.

An Air Force helicopter flight instructor named George Abbey was driving from his hometown of Seattle, Washington, to Randolph Air Force Base in San Antonio, Texas, when he learned of the Russian achievement. He knew his route would intersect Sputnik's orbital path the next day allowing him an opportunity to see it with his own eyes. A

---

[1] Craig McEwen, "West Fargo Astronaut Recalls 1985 Challenger Launch," *Grand Forks Herald*, July 29, 2015, https://www.grandforksherald.com/news/3807216-west-fargo-astronaut-recalls-1985-challenger-launch.

few minutes before it passed overhead, Abbey pulled over and stared into the sky. He was stunned watching the satellite streak overhead.[2]

The amazement turned into widespread fear of Russians controlling outer space. If they could orbit a satellite, the Cold War rivals could spy on the United States and even drop bombs.

Sputnik was the Soviet answer to a 1952 challenge from the International Council of Scientific Unions. It established July 1, 1957, to December 31, 1958, as the International Geophysical Year (IGY), expecting solar activities to peak. A 1954 resolution called for nations to develop Earth-orbiting satellites to study the planet and its atmosphere during this ideal timeframe.

It did not take America long to respond to Sputnik. The National Aeronautics and Space Administration was formed in 1958 and announced the goal of sending men into space. NASA's Space Task Group initially ran Project Mercury, America's first manned spaceflight program. The task group, led by Bob Gilruth, was located at Langley Research Center in Virginia. The Langley engineers were an accomplished group but eccentric compared to the by-the-book approach of the Air Force.

By 1960, George Abbey was working with the Air Force Dyna-Soar program, short for Dynamic Soaring. Dyna-Soar planned to launch a space plane from a rocket, perform various military functions, and land like an airplane. As the project moved beyond the drawing board stage, NASA arranged a joint conference with the Air Force to discuss some of the primary issues involved with spaceflight and reentry. Abbey was sent to Langley to observe.

The week before the conference, NASA engineers presented their material for peer review during rehearsals. George Abbey was shocked to see the heated free-for-all that ensued. Instead of the disciplined, cordial responses he was accustomed to in the Air Force, the unruly engineers in the audience were shouting and critiquing even the most minor points. Abbey was equally shocked at the end of the day when the same people were laughing and enjoying each other's company as if they were best of friends.

---

[2] Michael Cassutt, *The Astronaut Maker: How One Mysterious Engineer Ran Human Spaceflight for a Generation*, (Chicago: Chicago Review Press, 2018), 3–5.

The raucous interactions continued the next day. Abbey began to recognize some of the leading players, including Gilruth, George Low (head of manned spaceflight at NASA headquarters in Washington), and a lead engineer named Max Faget, who was instrumental in the design of the Mercury spacecraft. The following week during the official conference, all of the NASA presenters received no complaints about their revised material. Abbey was impressed at the benefit of immediate face-to-face peer review. It served him well in the years to come.

Ever since Russia embarrassed the United States by launching the first satellite, George Low and Bob Gilruth wanted the first man in space to be an American. Alan Shepard and his Mercury capsule were ready for flight in March 1961. Low and Gilruth wanted to launch then, but the cautious team working on the Redstone booster insisted on one more unmanned test flight. Shepard's flight was pushed back until May. During that delay, Low received a phone call at 2:00 a.m. on April 12; Yuri Gagarin was orbiting Earth in a Soviet spacecraft. The engineer called it one of his deepest disappointments. [3] Three weeks later, Shepard became the first American in space.

The Soviets were consistently a few steps ahead in the space race. Led by Chief Designer Sergei Korolev, whose identity was unknown to the public until he died in 1966, they assembled an impressive list of accomplishments. In 1957 they not only sent the first satellite into orbit but also launched the first living creature into space, a dog named Laika.

Two years later, they achieved three notable successes with their unmanned Luna probes. Luna 1 flew by the Moon, Luna 2 impacted the lunar surface, and Luna 3 photographed the far side of the Moon. [4] In 1960 the Soviets safely returned two dogs sent to space on Sputnik 5, paving the way for Yuri Gagarin to be the first man in space in 1961. Valentina Tereshkova became the first woman in space in 1963. Alexei Leonov performed the first space walk in 1965, and an unmanned Luna 9 probe successfully landed on the Moon the following year.

Yuri Gagarin's historic flight on April 12, 1961, was especially troubling to America's young President, John F. Kennedy. More than national pride was at stake. The nation was entrenched in geopolitical

---

[3] Richard Jurek, *The Ultimate Engineer: The Remarkable Life of NASA's Visionary Leader George M. Low*, (Lincoln: University of Nebraska Press, 2019), 55.
[4] Decades later, it was revealed the actual goal of Luna 1 was to impact the Moon.

tension with Russia, and space was emerging as a primary focal point. Each country sought to show superiority in military strength and technology. The nuclear arms race made it even more intense. The Soviets could reach the United States with an Intercontinental Ballistic Missile (ICBM), and it was important for President Kennedy to respond quickly and decisively.

On April 14, just two days after Gagarin's flight, a decisive meeting took place in Washington, DC. Ted Sorensen, White House counsel to President Kennedy, called this the most critical day in Kennedy's decision-making process to go to the Moon.[5] Just four men joined Sorensen in the meeting: Jerome Weisner (President's Science Advisory Committee), David Bell (Director of the US Office of Management and Budget), James Webb (NASA Administrator), and Hugh Dryden (NASA Deputy Administrator). They had different opinions regarding manned space exploration.

After many hours of debate, the consensus was a manned lunar landing was the best way for the United States to catch and surpass the Russians. It would involve next-generation technology, giving America's capitalist system and resources a chance to flex their muscles. The reality was the United States had never launched a man into space. Alan Shepard's *Freedom 7* sub-orbital flight was scheduled for the next month, and it had to go well.

There was something else on Kennedy's mind that 14th day of April. Cuban exiles trained by the United States military and Central Intelligence Agency assembled their invasion forces near the Bay of Pigs in Cuba. The ultimate goal was to overthrow the communist government of Fidel Castro. Air strikes began on April 15, and invasion forces landed on April 17. They were soundly defeated by the Cuban military trained by the Soviet Union and other Eastern Bloc nations.[6] The ensuing Cuban Missile Crisis was a confrontation between the world's two superpowers, the United States and the Soviet Union, and the closest they came to nuclear war. The tension between the two had been brewing since the end of World War II. It was known as the "Cold

---

[5] Charles Murray and Catherine Bly Cox, *Apollo: The Race to the Moon*, (New York: Simon and Schuster, 1989), 78.
[6] Murray and Cox, *Apollo*, 79–80.

War" since it did not involve large-scale fighting, but the future was very uncertain.

DECISION TO GO TO THE MOON

Kennedy suffered two humiliating defeats in one week, only three months after taking office as President. He looked to the space program to turn the tide. After Shepard's successful flight, it was clear that the United States had to commit itself to something bold. In Weisner's opinion, the decision was "a political, not a technical issue." It was "a use of technological means for political ends."[7]

On May 25, the President gave his *Special Message to Congress on Urgent National Needs.* He began by stating, "These are extraordinary times. And we face an extraordinary challenge. Our strength as well as our convictions have imposed upon this nation the role of leader in freedom's cause."[8]

Kennedy spoke of winning the crucial battle between freedom and tyranny. He explained the need to unleash the nation's talent and resources to achieve long-range goals. And then he spoke these memorable words:

> I believe that this nation should commit itself to achieving the goal, before this decade is out, of landing a man on the Moon and returning him safely to the Earth. No single space project in this period will be more impressive to mankind or more important for the long-range exploration of space, and none will be so difficult or expensive to accomplish. We propose to accelerate the development of the appropriate lunar spacecraft. We propose to develop alternate liquid and solid fuel boosters, much larger than any now being developed, until certain which is superior. We propose additional funds for other engine development and for unmanned explorations— explorations which are particularly important for one purpose which this nation will never overlook: the survival of the man

[7] John M. Logsdon, *The Decision to Go to the Moon: Project Apollo and the National Interest,* (Cambridge: MIT Press, 1970), 118.
[8] John F. Kennedy, "Address to Joint Session of Congress, May 25, 1961," JFK Library Foundation, accessed October 3, 2018, https://www.jfklibrary.org/learn/about-jfk/historic-speeches/address-to-joint-session-of-congress-may-25-1961.

who first makes this daring flight. But in a very real sense, it will not be one man going to the Moon—if we make this judgment affirmatively, it will be an entire nation. For all of us must work to put him there.[9]

This challenge pitted one country against another, one value system against another, one economic system against another, and liberty versus tyranny. President Kennedy boldly threw down the gauntlet. He was betting on freedom to succeed over Communism.

He not only inspired the country, but he also challenged her. Even though resources and talent abounded, the United States still needed to develop the procedures, material, and infrastructure to accomplish such a mission. Apollo flight controller William Reeves summarized the situation:

> At the time Kennedy made that speech in 1961, the centers [Houston, Huntsville, Cape Kennedy] did not exist as they do today, the vehicle designs did not exist, the methodology of how to get to the Moon and back did not exist, we had never flown an American in space except Alan Shepard for the one lob shot, we had never rendezvoused in space before. To solve all of those problems in the eight years between Kennedy's speech and the landing in 1969, we not only designed the vehicles, built the vehicles, and solved all of those technical problems, but we built the entire infrastructure of the agency. All the centers were put together, all the contracts were let, all of the contractors were hired, and all the government employees were hired and trained, so it was a monumental effort involving hundreds of thousands of people.[10]

Kennedy was realistic, recognizing that the Soviets would likely keep their lead for some time and continue to achieve impressive feats. The American response would be carried out in clear view of the watching world whether it resulted in success or failure. It would not be one man landing on the Moon but the entire nation.

---

[9] John F. Kennedy, "Address to Joint Session of Congress, May 25, 1961."
[10] William Reeves, discussion with author at the Apollo 11 50th Anniversary Luncheon, Houston, July 16, 2019.

HOW TO GET THERE

Following Kennedy's speech, NASA had to make enormous decisions regarding boosters, spacecraft, personnel, and procedures. But before confronting any of those issues, they had to decide how to get to the Moon.

NASA considered three strategies. The first was "Direct Ascent," similar to what many people read about in science-fiction stories. One colossal rocket would take off from Earth, land on the Moon, and return to Earth as one complete vehicle. A primary argument against this method was the amount of time it would take to develop such a massive spaceship.

The second proposal was "Earth Orbit Rendezvous" involving the rendezvous of smaller Saturn boosters in Earth orbit. From there, a lunar vehicle would be assembled and travel to and from the Moon. An argument for this method was the Saturn boosters were already far along in development.

Instead, NASA chose a third option, "Lunar Orbit Rendezvous" (LOR). This strategy emerged from NASA Langley Research Center in Hampton, Virginia. LOR called for the main spacecraft (Command Module) to orbit the Moon while a small landing craft (Lunar Module) would depart from it, land on the lunar surface, and later rendezvous with the Command Module, which would return to Earth.

The most vocal proponent for this method was Dr. John Houbolt, who argued it would reduce the complications inherent in the other proposals. He said, "the Lunar Orbit Rendezvous offered a chain-reaction simplification on all the back effects: development, testing, manufacturing, erection, countdown, flight operations, etc. All would be simplified."[11] Houbolt knew this was the most efficient way to achieve a lunar landing and zealously preached its merits to whoever would listen. His efforts and arguments eventually prevailed. At a press conference in July 1962, NASA announced its choice of LOR, a landmark decision in achieving the lunar landing by the end of the decade.

---

[11] Barton C. Hacker and James M. Grimwood, *On the Shoulders of Titans: A History of Project Gemini*, (Washington, DC: National Aeronautics and Space Administration, 1977), 15.

RICE UNIVERSITY

Once NASA decided on the Lunar Orbit Rendezvous method, the task at hand became clear. The President spoke at Rice University in Houston on September 12, 1962, addressing many of the challenges:

> This country was conquered by those who moved forward, and so will space.... The exploration of space will go ahead, whether we join in it or not, and it is one of the great adventures of all time, and no nation which expects to be the leader of other nations can expect to stay behind in the race for space.[12]

Kennedy vowed that America would not merely be a part of the space race but lead it, utilizing space for peace instead of war. He continued:

> We choose to go to the Moon. We choose to go to the Moon in this decade and do the other things, not because they are easy, but because they are hard, because that goal will serve to organize and measure the best of our energies and skills, because that challenge is one that we are willing to accept, one we are unwilling to postpone, and one which we intend to win, and the others, too.
>
> But if I were to say, my fellow citizens, that we shall send to the Moon, 240,000 miles away from the control station in Houston, a giant rocket more than 300 feet tall, the length of this football field, made of new metal alloys, some of which have not yet been invented, capable of standing heat and stresses several times more than have ever been experienced, fitted together with a precision better than the finest watch, carrying all the equipment needed for propulsion, guidance, control, communications, food and survival, on an untried mission, to an unknown celestial body, and then return it safely to Earth, re-entering the atmosphere at speeds of over 25,000 miles per hour, causing heat about half that of the temperature of the Sun—almost as hot as it is here today—and

---

[12] John F. Kennedy, "Address at Rice University on the Nation's Space Effort," September 12, 1962, JFK Library Foundation, https://www.jfklibrary.org/learn/about-jfk/historic-speeches/address-at-rice-university-on-the-nations-space-effort.

do all this, and do it right, and do it first before this decade is out—then we must be bold....

Many years ago, the great British explorer George Mallory, who was to die on Mount Everest, was asked why did he want to climb it. He said, "Because it is there." Well, space is there, and we're going to climb it, and the Moon and the planets are there, and new hopes for knowledge and peace are there. And therefore, as we set sail, we ask God's blessing on the most hazardous and dangerous and greatest adventure on which man has ever embarked.[13]

On November 22, 1963, President Kennedy was assassinated in Dallas, Texas. Many have said the blood of martyrs is the seed of the church. The same was true for Kennedy's vision. Gene Kranz said, "After his assassination, it moved from being a challenge to literally a crusade. This was now our mission to win this battle for President Kennedy. It was visceral. We are going to do it, and we are the right people to do it, and we are going to do it in the time frame he said we will do it."[14]

John Houbolt, an engineer at the Langley Research Center in Virginia, successfully lobbied for the Lunar Orbit Rendezvous strategy for Apollo. Photo credit: NASA.

[13] John F. Kennedy, "Address at Rice University on the Nation's Space Effort."
[14] Brendan W, "Failure is Not an Option, A Flight Control History of NASA," October 21, 2014, video, 1:29:45, https://www.youtube.com/watch?v=7f51Jzm7M4w.

President Kennedy (right) with Wernher von Braun. Photo credit: NASA.

President Kennedy looks inside Friendship 7 with John Glenn
after Glenn's orbital flight in February 1962. Photo credit: NASA.

2

PROGRAM MANAGER

FORMATION OF NASA

In the early days of World War I, it became apparent airplanes would be an effective tool in locating the enemy. Countries began to focus on aeronautics, the science of flight. The United States formed the National Advisory Committee for Aeronautics (NACA) in 1915, modeled after similar European agencies. For over 40 years, NACA made significant contributions to the research and development of aviation. During that time, NACA built advanced facilities including the Langley Research Center in Virginia, Ames Research Center in California, and the Lewis Research Center in Ohio.

The launch of Sputnik prompted a sense of urgency about the United States' role in space. President Dwight Eisenhower was determined the military would not run the effort. He submitted a proposal to Congress with the intention that the civilian NACA would serve as the roots of a new space agency. Congress modified the proposal and enacted the National Aeronautics and Space Act of 1958. According to General Sam Phillips, "That was a national response, political and otherwise, to the Sputnik accomplishment in October of 1957 by the USSR, and to some extent was an expression of dissatisfaction by our national leaders with the way in which the United States space programs were progressing."[1] The first declaration of the

---

[1] General Samuel C. Phillips, interview by Lt Col J.B. Kump for the U.S. Air Force Oral History Project, February 22, 1989. Part 4 of 6. Video graciously sent to author by Kathleen Phillips Esposito. Now available on YouTube,

Space Act states, "The Congress hereby declares that it is the policy of the United States that activities in space should be devoted to peaceful purposes for the benefit of all mankind."[2]

The Space Act of 1958 established a new organization, the National Aeronautics and Space Administration (NASA), and stated the NACA would "cease to exist" and all of its property, personnel, and resources would be transferred to NASA.[3] Also transferred to NASA were two components of the US Army's ballistic missile program, namely the Redstone Arsenal in Huntsville, Alabama, and the Jet Propulsion Laboratory developed by the California Institute of Technology (Caltech).

The Redstone Arsenal featured Wernher von Braun and his team of scientists from Germany. The Jet Propulsion Laboratory (JPL) was a contractor for the Army. Working closely together, the Redstone Arsenal and JPL launched the nation's first satellite, Explorer 1, on January 31, 1958. These two talented groups, along with the research centers at Langley, Ames, and Lewis provided NASA with a good foundation on which to build.

The space administration quickly grew. Von Braun's group in Huntsville developed into the Marshall Space Flight Center. Their engineers traveled to Florida to conduct launch activities at the Cape Canaveral Air Force Station. As launch complexes developed, the facility became the Launch Operations Center directed by Kurt Debus.

The Space Task Group, based out of Langley and led by Robert Gilruth, ran the Mercury program. After President Kennedy set the goal of a lunar landing by the end of the decade, a larger organization and more facilities were needed. Rice University donated land in Houston, and the task group moved there, becoming the Manned Spacecraft Center (MSC). NASA had tremendous technical capabilities, but since

---

https://www.youtube.com/watch?v=K2UiWEnMK-0&list=PLmYqE0mjDMS680GwV9k7wSfXpxHteBtKF&index=2.

[2] NASA, "The National Aeronautics and Space Act of 1958 (Unamended)," Sec. 102 (a), accessed October 4, 2020, https://history.nasa.gov/spaceact.html. The famous phrase "We Came in Peace for All Mankind" read from the lunar surface by Neil Armstrong and Buzz Aldrin summarizes the sentiment.

[3] NASA, "The National Aeronautics And Space Act of 1958," Sec. 202.(a) and Sec. 301.(a).

its formation consisted of various organizations pieced together, no coordination existed between the centers.

By 1963, the lack of production from the space agency caused frustration. They were behind schedule, significantly over budget, and people were losing hope of reaching the Moon by the decade's end. To remedy the situation, NASA Administrator James Webb appointed George Mueller (pronounced "Miller") to lead the Office of Manned Space Flight. Mueller created a structure where Directors of the three major centers reported directly to him, namely Gilruth at the Manned Spacecraft Center in Houston, von Braun at the Marshall Space Flight Center in Huntsville, and Debus at the Launch Operations Center at Cape Canaveral.[4]

Mueller also introduced a structure designed around programs, which at that time included Gemini, Apollo, and advanced developments, with these program offices reporting to him initially.[5] As the "acting" Program Director, Mueller needed someone who could flesh out the details and run it. He wanted to tap into the Air Force leadership experience and hire the best program manager he knew of—General Sam Phillips.[6]

Phillips was a fighter pilot in the Eighth Air Force during World War II. After the war, he played a vital role in developing the B-52 bomber, then became Program Manager for the Air Force Minuteman Intercontinental Ballistic Missile program.[7] Soon after the first flight of the Minuteman missile, Secretary of Defense Robert McNamara and his guest James Webb visited the Ballistic Missile Division in California. They received a briefing on the Minuteman program from Phillips and were impressed by how he ran the operation. General Phillips recalls,

> I gave Secretary McNamara a picture which I signed; it was the first flight and an unprecedented success. It was an "all-up

[4] George E. Mueller, interview by Summer Chick Bergen for Johnson Space Center Oral History Project, January 20, 1999, 3.
[5] General Sam Phillips, interview by Martin Collins, National Air and Space Museum, September 28, 1989, Tape 1, Side 1, accessed July 13, 2020, https://airandspace.si.edu/research/projects/oral-histories/TRANSCPT/PHILLIP6.HTM.
[6] George Mueller, interview by Summer Chick Bergen, January 20, 1999, 3.
[7] The program was named after the Colonial soldiers known for being ready to fight on very short notice during the Revolutionary War.

mission." In other words, although it was launched from a surface pad, the missile was complete. It had all the stages, guidance and reentry vehicle, and a target out near Ascension Island, and the impact was well within our specifications.[8]

The Minuteman program reached the stage of flight testing on schedule and budget. Phillips decided the program would go operational in October 1962 during the Cuban Missile Crisis, giving the United States a retaliatory weapon in case Russia attacked American cities. His decisiveness and ability to successfully lead a complicated program earned the attention of George Mueller. Webb and Mueller knew he would be the perfect man to run Apollo, but it would be difficult convincing the Air Force to loan one of their top managers to a civilian organization.

James Webb broached the subject with McNamara. Then conversations involving Webb, Mueller, General Eugene Zuckert (Air Force Secretary), General Curtis LeMay (Air Force Chief of Staff), and General Bernard Schriever (Commander of the Air Force Ballistic Missile Division) eventually led to Phillips' assignment to NASA on detached duty, to run the Apollo Program Office in Washington.[9]

As Vice Commander of the Ballistic Missile Division and looking at a promotion soon, Phillips was not sure going to a civilian organization would be a good career move, thinking he would be forgotten at NASA. Although some at the space agency had a wait-and-see attitude, they soon warmed up to Phillips' humility, hard work, and strong gifts. By the fall of 1964, less than nine months after he began at NASA, it was obvious things were running well. Mueller removed his name as the "acting" Program Director on the organizational chart, replacing it with Phillips, who kept the title until his mission was accomplished with the successful flight of Apollo 11.[10] Phillips' management abilities were invaluable to NASA as he balanced Apollo's results with its schedule and cost.

---

[8] General Sam Phillips, interview by J.B. Kump.
[9] General Sam Phillips, interview by Martin Collins, September 28, 1989, Tape 1, Side 1.
[10] General Sam Phillips, interview by Martin Collins, September 28, 1989, Tape 1, Side 1.

He reported to Mueller, who answered directly to NASA Administrator Webb. It did not take Phillips long to size up the situation. "NASA had developed to be a very, very professional technical organization, but they had almost no management capability nor experience in planning and managing large programs. They had tremendous technical competence and depth, but they had very little experience and people with experience in management."[11]

Phillips knew precisely where to find the help he needed. He went to LeMay, Zuckert, and Schriever, asking for assistance. "We wound up in a matter of days with 50 Air Force officers assigned in key places in the NASA centers and elsewhere in the agency," remembers Phillips in a roundtable discussion. Schriever, also in the conversation, quickly added, "Sam got even with me. I made him available to NASA; then he robbed me!"[12]

One hundred more Air Force officers followed by the end of 1964, including George Abbey. When the Air Force canceled their Dyna-Soar program, Abbey received orders to join NASA. He was raised in Oregon and did not like the Texas heat and humidity. He contacted friends at Langley in Virginia working on the Lunar Orbiter project and asked them to write General Phillips requesting permission for Abbey to help there. Phillips replied, "I'm very pleased to hear all that Captain Abbey is doing to support the Lunar Orbiter Program. He has thirty days to complete all of those activities and report to Houston."[13]

CONFIGURATION CONTROL BOARDS

Integral to Sam Phillips' leadership was the concept of Configuration Management, limiting the endless design changes which typically derail budgets and schedules. It was an essential ingredient in controlling the Minuteman program, and Phillips employed the same system throughout NASA. As the Apollo Program Director, he introduced a structure where each level of the hierarchy implemented Configuration Control Boards (CCBs) run by Program Managers. These men were

---

[11] Richard H. Kohn, *Reflections on Research and Development in the United States Air Force*, (Washington, DC: Center for Air Force History, 1993), 87.
[12] Richard Kohn, *Reflections on Research and Development in the US Air Force*, 87.
[13] Cassutt, *The Astronaut Maker*, 33.

decision-makers. After hearing all of the necessary input, they made the final decisions on the spot. The notion of "Let's think about it and decide next time" had no place in the CCB.

The Configuration Control Board meetings were serious business. An influential Configuration Board existed in Houston, where the Apollo Spacecraft Program Office (ASPO) was located. By 1967, George Low was the ASPO Program Manager, serving as Chairman of the meetings and making the final decisions.

Apollo Flight Director Gerry Griffin recalls, "The CCB was *effective*; it was hard to make changes." That was in stark contrast to the early days of NASA when astronauts would go to McDonnell Aircraft Corporation, which built the Mercury and Gemini spacecraft. The contractor wanted to please these national heroes, so any changes they suggested, McDonnell would implement. Altering one thing often had unintended consequences affecting other systems of the spacecraft. The result could easily add weight, increase costs, or delay the schedule. Things got out of hand until the CCB was established. Griffin continues, "When you went to the CCB with a proposed change, you'd better be ready because you were going to get *grilled:* 'Why do you need it? What does it do for you? Is it a safety of flight issue or just a convenience?'"[14]

Configuration Management Offices were placed in each center to ensure the implementation of the method. The same decisive structure was placed throughout NASA, extending even to the contractors associated with Apollo. Phillips required firm specifications and engineering requirements that defined and documented the objectives of a design. Once a preliminary design review approved specifications, any future changes had to go through the Configuration Control Board. Likewise, once a piece of hardware passed the First Article Configuration Inspection, any future changes had to go through the CCB. As George Mueller explained to Wernher von Braun, Configuration Management doesn't mean you can't change a design; it means those changes are documented and communicated.[15] Phillips' system also called for acceptance testing at the contractors' facility,

---

[14] Gerry Griffin, interview by author, July 13, 2020.
[15] Stephen B. Johnson, *The Secret of Apollo: Systems Management in American and European Space Programs*, (Baltimore: JHUP, 2006), 139–140.

inspecting the hardware to ensure it met requirements before shipping to NASA.[16]

These Configuration Management Offices were strategic locations for many of Low's Air Force officers who understood the tiered system and everyone's roles. And NASA was glad to have them. "In the early going, if we didn't have the military guys in the Mission Control Center, we would have been hurting," said Gerry Griffin. He added:

> We had Air Force officers in Mission Control and all over MSC, particularly in early Apollo. None of them wore uniforms. Those military guys, boy they were good. A number of our network guys were military people because they had tracking stations long before NASA got involved with them. They knew how the facilities around the world worked, and how that data had to flow all over the country, so they were really, really good at that. We used all the assets we could get our hands on.[17]

Houston did not need Washington to tell them how to design or fly a spacecraft, Huntsville did not need Washington to tell them how to build a Saturn launch vehicle, and the Cape did not need Washington to tell them how to launch it. But they needed the management structure to ensure their immense technical skills bore fruit.[18]

The system was beneficial to organizations outside NASA centers. Phillips did not hesitate to get involved with contractors when he spotted a problem. By 1965 NASA was concerned about North American Aviation, manufacturer of the S-II second stage of the Saturn launch vehicle, as well as the Command and Service Modules. After an S-II stage failure during a structural test, engineers at the Marshall Space Flight Center in Huntsville organized a team to evaluate the situation at North American. They reported their assessment to von Braun, saying, "The S-II program is out of control.... It is apparent that management of the project at both the program level and division level has not been

[16] NASA, "Apollo Configuration Management Manual," (Washington, DC: National Aeronautics and Space Administration, 1970), Section 5.2.1.6, 8, https://ntrs.nasa.gov/citations/19800071127.
[17] Gerry Griffin, interview by author, August 9, 2020.
[18] General Sam Phillips, interview by Martin Collins, September 28, 1989.

effective…. In addition to the management problems, there are still significant technical difficulties with the S-II stage."[19]

General Phillips led a special "Tiger Team" of NASA engineers to Downey, California, where they performed a comprehensive examination of North American Aviation (NAA).[20] Their investigation resulted in a blistering report by Phillips outlining the problems throughout the organization. The "Phillips Report" was thorough. Its first summary finding:

> NAA performance on both programs is characterized by continued failure to meet committed schedule dates with required technical performance and within costs. There is no evidence of current improvement in NAA's management of these programs of the magnitude required to give confidence that NAA performance will improve at the rate required to meet established Apollo program objectives.[21]

North American responded by selecting a retired Air Force General, Robert Greer, to lead the S-II program. Greer previously served as Chief of Staff for the Air Force Guided Missiles program and General Schriever's Ballistic Missile Division. He implemented many procedures from his Air Force experience, improving the S-II program.

ALL-UP TESTING

An aggressive testing strategy was another way to control the schedule and cost. George Mueller and Sam Phillips were firm believers in "all-up testing," evaluating all flight components simultaneously. Von Braun's team wanted to take a conservative incremental approach, with only one significant change per flight. They planned to fly the first stage with only a dummy second and third stage. Then once the first stage was certified, they would add an active second stage, following that pattern until testing a complete launch vehicle with operational Command and

---

[19] Roger E. Bilstein, *Stages to Saturn, A Technological History of the Apollo/Saturn Launch Vehicles*, (Washington, DC: National Aeronautics and Space Administration, 1980), 224.
[20] "Tiger Team" was an Air Force term for a group of specialized troubleshooters.
[21] NASA, "The Phillips Report, 1965–1966," NASA Historical Reference Collection, accessed July 13, 2020, https://www.history.nasa.gov/Apollo204/phillip2.html.

Service Modules. Such a sequence would require four flights in the best-case scenario, plus the pre-launch preparation and post-flight analysis, adding at least a year to the Saturn program.

The all-up strategy would test every stage at once. It was risky, gambling that all the extensive ground tests of each component would have detected flaws. If successful, it would leapfrog several expensive and time-consuming steps. Mueller and Phillips had seen all-up testing keep the Minuteman on schedule while avoiding the added costs of extra test flights. Mueller said, "It was pretty clear that there was no way of getting from where we were to where we wanted to be unless we did some drastically different things, one of which was all-up testing."[22]

Wernher von Braun was startled at Mueller's approach. Eventually, the conservative rocketeer was convinced Mueller was right. "It sounded reckless, but George Mueller's reasoning was impeccable," said von Braun. "In retrospect, it is clear that without all-up testing, the first manned lunar landing could not have taken place as early as 1969."[23]

The media and public often overlook the importance of George Mueller and Sam Phillips, although they were highly regarded by their peers. Phillips never understood why history did not give Mueller the credit he deserved for Apollo's success.[24] The same is often said about Phillips. On the 50th anniversary of the launch of Apollo 11, the Public Affairs Office of the Space and Missile Systems Center in Los Angeles paid tribute to General Phillips, including the following account:

> At a small dinner party before the Apollo 10 dress rehearsal mission in May 1969, Dr. Wernher von Braun, director of NASA's Marshall Space Flight Center in Huntsville, Alabama, singled out Phillips as the man to whom the greatest credit belonged for pulling the many pieces of the Apollo program together and making them work, on time and within budget.[25]

---

[22] Bilstein, *Stages to Saturn*, 349.
[23] Edgar M. Cortright, ed, *Apollo Expeditions to the Moon*, (Washington, DC: National Aeronautics and Space Administration, 1975), 50.
[24] Murray and Cox, *Apollo*, 159.
[25] Space and Missile Systems Center, "Gen. Samuel C. Phillips: A retrospective on the director of the Apollo program and former commander of the Space and Missile Systems Organization," https://www.losangeles.spaceforce.mil/News/Article-

George Mueller (left) and Sam Phillips brought US Air Force management structures to NASA, enabling the space agency to meet its end-of-decade deadline. Photo credit: NASA.

Apollo 11 launch, from left to right: Charles Matthews, Wernher von Braun, George Mueller, Sam Phillips. Photo credit: NASA.

Display/Article/1917223/gen-samuel-c-phillips-a-retrospective-on-the-director-of-the-apollo-program-and/.

3

USS MOLLY BROWN

GEMINI III

In the early 1960s, the Mercury, Gemini, and Apollo programs were in motion concurrently, each with distinct objectives. The final Mercury flight occurred in May 1963 with Gordon Cooper's *Faith 7,* which completed 22 orbits in 34 hours. Mercury successfully fulfilled its goals by demonstrating man was capable of orbiting Earth, functioning in space, and returning safely. The program also solidified the core elements of flight control. However, there was still a long way to go before catching the Russians, not to mention meeting President Kennedy's lunar deadline.

NASA created Gemini to develop and test systems and procedures required for lunar flights, such as rendezvous, docking, Extra Vehicular Activities (EVAs), and long-duration periods in space.[1] Rendezvous is the complicated process of bringing two orbiting spacecraft together. It was something the Russians had never accomplished and was essential for getting two moonwalkers from the landing module to the Apollo Command Module for the return flight to Earth.

After astronaut Gus Grissom flew the second manned Mercury flight in *Liberty Bell 7,* he began working intensely with the engineers

---

[1] EVAs are activities performed outside the spacecraft by astronauts wearing space suits. In Project Gemini, they were "spacewalks" in Earth orbit. In Apollo, EVAs included moonwalks by the Commander and Lunar Module Pilot. During the last three lunar missions, Command Module Pilots performed deep-space EVAs to retrieve film canisters from the Service Module on the way back to Earth.

designing the Gemini spacecraft. Like its Mercury predecessor, the Gemini capsules were constructed by McDonnell Aircraft Corporation in St. Louis, Missouri. Grissom gave the McDonnell engineers valuable feedback throughout the design process.

The Gemini capsule carried two men, was twice as heavy as the Mercury capsule, and required a more powerful booster. The Titan II fit the bill and became the launch vehicle for all Gemini flights. It was a two-stage liquid-fueled rocket, modified from the Titan II Intercontinental Ballistic Missile, and provided a smoother ride than the Atlas booster did for Mercury orbital flights.

Grissom was highly motivated. A problem with the hatch on *Liberty Bell 7* caused it to open prematurely after splashdown. The astronaut escaped, but he nearly drowned as the recovery helicopter struggled to lift the waterlogged capsule from the Atlantic Ocean. *Liberty Bell 7* rapidly filled with water and the added weight put too much stress on the helicopter's engine. The pilot cut it loose, and the spacecraft sank to the bottom of the ocean. Another helicopter flew in to rescue Grissom.[2]

The incident placed a dark cloud over an otherwise successful mission. Immediately there was speculation Grissom panicked and blew the hatch himself. In their excellent volume *Into That Silent Sea: Trailblazers of the Space Era, 1961–1965,* Francis French and Colin Burgess go into great detail about the episode.[3]

The explosive hatch was supposed to be triggered when the helicopter's cable was attached to a recovery loop on the capsule. According to fellow Mercury astronaut Deke Slayton and NASA engineer Sam Beddingfield, if Grissom manually opened the hatch, he would have struck the plunger with so much force his hand would have suffered a deep and visible bone bruise.

At the end of a later Mercury flight, Wally Schirra had to manually open his hatch by hitting the plunger. He quickly showed the resulting

---

[2] *Liberty Bell 7* was found at the bottom of the Atlantic Ocean by Curt Newport on a 1999 expedition funded by the Discovery Channel. It was the deepest commercial salvage project in history. The spacecraft has been restored and is permanently displayed at the Cosmosphere in Hutchinson, Kansas.

[3] Francis French and Colin Burgess, *Into That Silent Sea: Trailblazers of the Space Era, 1961–1965*, (Lincoln: University of Nebraska Press, 2007), 82–98.

bruise and cut on his hand in defense of his much-maligned friend. Grissom suffered no such injury when his hatch blew prematurely.

Gus Grissom focused his efforts on the development of the Gemini spacecraft. His fingerprints were all over it, to the point other astronauts called it the "Gusmobile." It was no surprise when Grissom was named Commander of the first manned Gemini flight. He would sit in the left-hand "Command Pilot" seat. John Young was chosen to be in the right-hand "Pilot" seat.[4]

John Young was the perfect match for Grissom. Both possessed a sharp engineering mind, a dry wit, and impeccable flying credentials. Grissom had a degree in mechanical engineering from Purdue University, while Young had an Aeronautical Engineering degree from the Georgia Institute of Technology. Grissom flew over 100 combat missions during the Korean War and was an Air Force test pilot. Young was a test pilot for the US Navy and set two world records flying an F-4 Phantom II. From a standing start, he climbed to 3,000 meters (9,843 feet) in 34.5 seconds and 25,000 meters (82,021 feet) in 228 seconds as part of "Project High Jump." Grissom was selected as one of the Original Seven Mercury astronauts in 1959. Young was chosen for Astronaut Group 2 in 1962 and was the first from that class to fly in space. Mike Collins writes,

> The crew of Gemini III, Gus Grissom and John Young, were a matched pair. They were both good engineers who understood their machines and liked fooling with them. They were uncomfortable with the invasion of privacy the space program had brought into their lives and tried as hard as they could to deflect questions from themselves to their beloved machines. They were generally taciturn, but both had strong opinions that could flash unexpectedly and, in Gus's case at least, angrily. Neither was interested in small talk, and they would endure uncomfortable silences rather than fill the void with what they considered ancillary trivia.... Both had a strong work ethic and kept long hours.[5]

---

[4] NASA used this terminology from the US Air Force and avoided the term "Copilot."
[5] Michael Collins, *Liftoff: The Story of America's Adventure in Space*, (New York: Grove Press, 1988), 83.

Dr. Bob Gilruth, Director of the Manned Spacecraft Center, made the official announcement of the prime crew on April 13, 1964. "In Project Mercury, a lot of improvements were made after the spacecraft got to the Cape, but with Gemini, most of the wrinkles were ironed out at the McDonnell factory in St. Louis," said Young. "Things looked so good even then that Gus went so far as to predict a launch date to me over a year ago. He pegged March 15, 1965. I thought he was off his rocker at the time but it turns out he was only a week off."[6] Grissom and Young spent the subsequent year testing and retesting the spacecraft for its initial shakedown cruise.

The crew named their capsule *Molly Brown*, a reference to a popular Broadway musical, *The Unsinkable Molly Brown,* based on a survivor of the Titanic tragedy. Not everyone at NASA was pleased with such an undignified name. "Some officials were a bit hesitant to go along with *Molly Brown*," said Grissom. "They asked me if I had a second choice, and I said, 'Sure, the *Titanic.*'" The officials relented, but it was the last time a crew was allowed to name their spacecraft until Apollo 9 when a Command Module flew with a Lunar Module. Each needed a call sign to distinguish them in communications.

Grissom and Young were determined this mission would go well. It would be a short flight, lasting only three orbits due to the geographical limitations of the worldwide tracking network.[7] The new Gemini capabilities necessitated extensive upgrades and testing of all locations in the network. In the South Texas coastal city of Corpus Christi, Tracking Station 16, known as "TEX," was undergoing a five million dollar expansion of the equipment used to track Mercury capsules. During the Mercury flights, its primary function was to report on the spacecraft's position. The upgrades converted it to be one of the first complete command and control stations. New equipment allowed TEX to not only talk to astronauts and monitor telemetry, but also take control of the Gemini spacecraft and bring it down out of orbit in the event of an emergency.[8] Training in Corpus Christi was well underway

---

[6] Gus Grissom and John Young, "Molly Brown was OK from the first time we met her," *Life Magazine*, April 2, 1965, Vol. 58, No. 13, 41.
[7] Hacker and Grimwood, *On the Shoulders of Titans*, 228.
[8] Bill Walraven, "Rodd Tracking Site Due Top Prominence," *Corpus Christi Caller Times*, June 5, 1965.

by the summer of 1964 when astronauts Neil Armstrong and Pete Conrad paid a visit in preparation for their upcoming duties as Capsule Communicators at remote tracking stations.[9]

Two initial unmanned Gemini flights went well, setting the stage for Gus Grissom and John Young to lead NASA into a new era. A primary aim of the mission was to use the Orbital Attitude and Maneuver System (OAMS) to alter the spacecraft's orbit, a necessity for future rendezvous missions.

On March 23, 1965, Cosmonaut Alexei Leonov and his crewmate Pavel Belyayev received a hero's welcome in Moscow five days after Leonov performed the first space walk in history. On the same day, Gemini III launched from Cape Kennedy. While Leonov's feat appeared to put the Soviets even farther ahead in the Space Race, the Gemini mission was about to demonstrate maneuvers vital to a lunar flight.

As planned, the Gemini mission lasted three orbits. John Young was awestruck by the sight outside his window. "There aren't enough words in the English language to describe the beauty," he said after the flight. "I was supposed to monitor the inertial guidance system... but it's just a tremendous effort to get your head back in the cockpit.... I was impressed, I'll tell you that."[10]

As they flew over Texas near the end of the first revolution (orbit), Grissom fired two thrusters for seventy-five seconds to change their elliptical orbit to a circular one. This OAMS burn "was more than just successful," say French and Burgess. "For NASA it was an outstanding triumph. For the first time ever, the flight path of a spacecraft had been manually altered by a pilot in orbit."[11]

Gemini had to show a spacecraft could adjust its orbital altitude and plane for the Apollo missions to rendezvous and achieve a lunar landing. Flight Director Chris Kraft said, "We'd just crossed a major milestone on the way to the Moon."[12] It happened under the watchful eye of the

---

[9] Spencer Pearson, "Astronauts Train Here on Tracking," *Corpus Christi Caller Times*, September 17, 1964. Both Armstrong and Conrad, who performed the first two lunar landings, did their Naval flight training in Corpus Christi, Texas.

[10] "Gemini: It Didn't Last Long Enough," *Newsweek*, April 5, 1965, Vol. LXV, No. 14, 50.

[11] Francis French and Colin Burgess, *In the Shadow of the Moon: A Challenging Journey to Tranquility, 1965–1969*, (Lincoln: University of Nebraska Press, 2007), 12.

[12] French and Burgess, *In the Shadow of the Moon*, 12.

tracking station in Corpus Christi. The TEX flight controllers took great pride that this historic orbit adjustment occurred overhead, calling it "the burn over the station."[13]

There were two more OAMS burns during the flight. On the second orbit, Grissom altered the orbital plane, twisting it slightly, demonstrating the maneuvering capacity of *Molly Brown*. The crew enjoyed their flight immensely but focused on the business at hand. Most of their conversation consisted of onboard technical exchanges, not using the air-to-ground loop. Their lack of communication with the ground drew the attention of Public Affairs Officer Paul Haney, who added mission commentary for the media so they could better understand the proceedings. Haney remarked:

> *Characteristically very little conversation going back and forth, strictly pilot talk, evaluating systems. Even Gordon Cooper, who was not noted as a loquacious pilot during his 22-orbit flight, has been moved to comment on the lack of comment by Gus Grissom and John Young during this flight. They are entirely satisfied with the operation of the spacecraft and our ground systems. We are two hours and fifty-seven minutes into this flight. During the Hawaii pass, Neil Armstrong, a fellow astronaut, confirmed the flight is "Go" for the third orbit.[14]*

Late in the flight, a two-and-a-half-minute OAMS burn over Texas brought the low point of the final orbit to 45 miles above Earth, low enough to assure reentry even if the retro-rockets failed. The retro-rockets worked as planned, so the crew met another objective by evaluating the controlled flight path entry. The Gemini capsule was designed with its center of gravity offset to one side, thereby creating lift as it entered Earth's atmosphere. By rotating the spacecraft, Grissom successfully used this capability to aim it toward the landing point.

The spacecraft splashed down with a severe jolt; *Molly Brown* lived up to her name and did not sink. "It was a great spacecraft, but it was no

---

[13] Lynn Pentony, "Men Here Will Guard Life of Men Up There," *Corpus Christi Caller Times,* March 21, 1965.

[14] Mission commentary by the Public Affairs Officers are indicated by italics. During Mercury and early Gemini flights, CAPCOMs were located at remote tracking stations.

boat," said Young.[15] Both men quickly became nauseous from the cockpit heat and bobbing on the waves. Young was used to it from the Navy, but Grissom came from an Air Force background and became seasick. "We lost several pounds waiting in that hot spacecraft until we were picked up," wrote Young.[16]

The Captain is usually the last to leave the ship, but Grissom wasted no time getting out when the recovery team arrived. Young teased him in front of the press, saying, "That's the first time I ever heard of a skipper leaving the ship first."

Grissom responded, "Well, I made you captain when I left."

Young said, "So now I was the captain of a Navy vessel for a few minutes, and I renamed it the *USS Molly Brown.*"[17]

The mission was a great success, a test pilot's dream. The crew had significant input on the design and development of the spacecraft and saw the fruit in actual flight. It went so well that Grissom let Young fly *Molly Brown* for a few minutes. Young said, "I thought I'd have to break his arm to take over even for that long."[18]

Gus Grissom and John Young thoroughly enjoyed the experience. Following the mission, they spent several days going through their technical debrief and listening to mission tapes. The pair had a blast, laughing at the many humorous moments from the flight.[19] Then they embarked on a public relations tour, including a week in Washington where they became the first astronauts to address Congress as they lobbied for funding needed to go to the Moon.[20]

One of the more memorable events involved a corned beef sandwich from a local deli purchased by Wally Schirra two days before the flight. He gave it to John Young, who hid it in a spacesuit pocket. One of the experiments on Young's slate was trying dehydrated space food contained in plastic bags. Water had to be applied from a water gun to make the meals edible. "I was sitting over there flying and John was fooling around with the food, and all of a sudden he asked me if I wanted

---

[15] John W. Young, *Forever Young: A Life of Adventure in Air and Space*, with James R. Hansen, (Gainesville: University Press of Florida, 2012), 82.
[16] Young and Hansen, *Forever Young*, 83.
[17] Grissom and Young, *Life Magazine*, 42.
[18] Grissom and Young, *Life Magazine*, 42.
[19] Young and Hansen, *Forever Young*, 83.
[20] Young and Hansen, *Forever Young*, 83.

to eat," explained Grissom. "Then he handed me a corned beef sandwich. I took a bite, but crumbs of rye bread started floating all around the cabin so I stowed the rest of it. I did try some of John's applesauce."[21]

After the flight, some members of Congress were not amused when they learned about it. George Mueller, Associate Administrator for Manned Space Flight, reassured them, "We have taken steps to prevent recurrence of corned beef sandwiches in future flights."[22]

Later Gemini missions also accomplished primary goals necessary for Apollo's lunar flights:

Gemini IV: As Commander Jim McDivitt looked on, Ed White became the first American to perform a spacewalk outside the capsule, testing the ability to perform an Extra Vehicular Activity.

Gemini V: Gordon Cooper and Pete Conrad completed an eight-day endurance mission in the cramped cabin to replicate the length of a lunar mission. Gemini V introduced fuel cells as the primary source of electrical power. A chemical reaction of liquid hydrogen and oxygen generated electrical power, with water being a byproduct for cooling equipment and drinking.

Gemini VII/VI: The joint mission performed the first rendezvous in space. Gemini VII launched first with Frank Borman and Jim Lovell, extending the endurance record to two weeks which was longer than any planned Apollo mission. Gemini VI, with Wally Schirra and Tom Stafford, launched 11 days later and performed a tricky rendezvous, placing the two spacecraft within one foot of each other. NASA had finally beaten the Russians to a critical milestone on the way to the Moon. "At the conclusion of the first space rendezvous, everyone in the Control Center waved their flags

---

[21] Grissom and Young, *Life Magazine*, 42.
[22] Collins, *Liftoff*, 85.

and showed how proud we were to be Americans," said flight controller Jerry Bostick. "I believe this was the first and last time we ever celebrated during a mission."[23]

Gemini VIII: Neil Armstrong and Dave Scott chased down an unmanned Agena target vehicle, and Armstrong performed the first docking maneuver in space. Surprisingly, the docked vehicles began to roll a few minutes later. Thinking the problem was with the Agena, Armstrong undocked, but the Gemini spacecraft began to spin wildly. A short circuit in the wiring caused a thruster to fire continuously, and the crew found themselves in a life-threatening situation. With blurred vision, they shut down all of the maneuvering thrusters. Then Armstrong used a separate reentry control system to stabilize the tumbling spacecraft. The crew survived due to their quick thinking and pilot skills but had to end their mission prematurely.

This turn of events caught the contingency recovery forces by surprise. Enjoying St. Patrick's Day festivities at a bar, they were called in for a rescue operation and thought it was a drill until they were over the spacecraft. In Houston, Flight Dynamics Officer Jerry Bostick heard the routine report of two recovery swimmers in the water tending to the spacecraft. Then he was alarmed to hear a third join them. The third swimmer was always a doctor, only deployed in the event of a medical situation. Several in Mission Control thought something must be wrong with the crew, but it turned out the doctor was taking care of the first two rescue swimmers who were getting sick. After everyone managed to board the recovery ship, Armstrong shook the rescue team members who were stretched out on the floor. "I think you're supposed to be checking us out medically," said the astronaut. He and Scott had to assist them through the process.[24]

---

[23] Jerry Bostick, "Trench Memories," in *From the Trench of Mission Control to the Craters of the Moon*, (CreateSpace Publishing, 2012), 162.
[24] Jerry Bostick, *The Kid from Golden: From the Cotton Fields of Mississippi to NASA Mission Control and Beyond*, (iUniverse Publishing, 2016), 99.

Gemini IX: After a failed Agena launch, Gemini IX performed three rendezvous with a backup vehicle called an Augmented Target Docking Adapter. Unfortunately, they could not dock because a shroud covering its docking collar had not completely opened. Later in the flight, Gene Cernan performed an exhausting two-hour EVA, fighting against a stiff space suit. Insufficient leverage caused an opposite reaction to every slight motion he attempted. Overheated, Cernan lost over 13 pounds, fogged up his visor, and barely got back inside the cramped spacecraft.[25] Flight Director Gene Kranz called the tense EVA "as tough as it ever got in the Mission Control Center."[26] Only three more Gemini missions remained, and the US had yet to conquer the challenges of operating outside the spacecraft.

A cutaway view of the two-man Gemini spacecraft.
Credit: NASA S-65-893.

---

[25] French and Burgess, *In the Shadow of the Moon*, 101.
[26] Gene Kranz, *Failure Is Not an Option: Mission Control from Mercury to Apollo 13 and Beyond*, (New York: Berkley Books, 2000), 186.

*John Young (left) and Gus Grissom were chosen to fly
the first manned Gemini flight. Photo credit: NASA.*

*Young and Grissom examine the development of the Gemini spacecraft.
Photo credit: NASA.*

John Young (left) and Gus Grissom in Molly Brown. Photo credit: NASA.

Dave Scott (left) and Neil Armstrong in their Gemini VIII capsule
waiting with swimmers until the recovery ship arrives. Photo credit: NASA.

4

THEY'VE DONE A LOT OF WORK GETTING READY FOR THIS ONE

GEMINI X

One of the mysteries of NASA is how Deke Slayton selected flight crews for the Gemini and Apollo missions. Slayton, a WWII bomber pilot and one of the Original 7 Mercury astronauts, was grounded for an erratic heartbeat before he ever flew in space. He then managed the astronaut office before being named Director of Flight Crew Operations. As 1965 gave way to 1966, a pattern emerged in the Gemini crew assignments. The backup crew of one mission became the prime crew three flights later.

Accordingly, Gemini VII backups Ed White and Michael Collins would fly on Gemini X. Those plans were altered on January 24, 1966, when NASA announced White was joining Gus Grissom and Roger Chaffee on the prime crew for the first manned Apollo mission. They also announced John Young as the new Commander of Gemini X, replacing White.

Young and Collins began intense training for their flight scheduled to launch in less than six months. It would be the most complicated mission yet attempted, with two rendezvous with two different Agenas, a docking, two EVAs, and over a dozen experiments. In addition, Collins was to perform enough onboard navigational exercises to make

Young start calling him "Magellan."[1] NASA was trying to gain as much experience as possible before Apollo flights began.

The Gemini flight schedule of 1966 was rocked by the fatal plane crash of astronauts Elliott See and Charlie Bassett, the prime crew for Gemini IX. They and backup crew members Tom Stafford and Gene Cernan flew to St. Louis on February 28 to inspect the spacecraft at the McDonnell plant and practice in their rendezvous simulator. See and Bassett flew in the first T-38, with Stafford and Cernan in the second.

The weather at Lambert Field in St. Louis was treacherous, with snow flurries mixed with rain and fog. Both planes aborted their first landing attempt. Stafford climbed back into the clouds to try another instrument approach, while See decided to stay under the cloud cover to keep the field in sight. Circling to his left, See clipped the roof of the very McDonnell building where technicians were putting the final touches on their spacecraft. The plane exploded and came to rest in a nearby parking area. Both men died instantly.

John Young was given the responsibility of telling Elliott See's wife, Marilyn, before she heard about it through the media. He called Marilyn Lovell, a close friend of Marilyn See, and asked her to be at the See home when he arrived. When Young knocked on the door, See's wife immediately realized something had happened. Young informed her of the accident, and Marilyn Lovell comforted the young widow.[2]

Elliott See was 38 years old; he and Marilyn had three children. See was a member of the second astronaut class, selected in 1962. Known as "The New 9," the class contained See, along with John Young, Neil Armstrong, Frank Borman, Pete Conrad, Jim Lovell, Jim McDivitt, Tom Stafford, and Ed White. All of the surviving members of his class played significant roles in the Apollo program.

Charlie Bassett was married with two children. He was a member of the third astronaut class, selected in 1963, with Mike Collins, Buzz Aldrin, Bill Anders, Alan Bean, Gene Cernan, Roger Chaffee, Walt Cunningham, Donn Eisele, Ted Freeman, Dick Gordon, Rusty Schweickart, Dave Scott, and C.C. Williams. Bassett was selected for the prime crew of Gemini IX, and Deke Slayton also assigned him to the

---

[1] Collins, *Carrying the Fire*, 171.
[2] Young and Hansen, *Forever Young*, 86–87.

backup crew of the second manned Apollo flight alongside Frank Borman and Bill Anders. Borman and Anders flew to the Moon on Apollo 8 with Jim Lovell.

The tragic accident has an interesting historical footnote. Most backup crews could look forward to having a prime crew assignment three missions later. But the Gemini X backup crew assignment was a dead-end job since the program would finish with Gemini XII. However, because of the accident, Gemini IX's backup crew of Stafford and Cernan became the prime crew. The Gemini X backup crew of Jim Lovell and Buzz Aldrin was reassigned to the Gemini IX backup crew and rotated to the prime crew of Gemini XII, flying the final Gemini mission. That was not as significant for Lovell, who spent two weeks in space aboard Gemini VII. But for Aldrin, it made all the difference in the world. Gemini XII was his first flight, and it is doubtful he would have been selected for the historic Apollo 11 crew if he had not flown a previous mission.

John Young and Mike Collins pressed on with training for their Gemini X flight. One day Slayton asked both Young and Collins to participate on the selection board of the next group of astronauts to be hired by NASA. Deke Slayton presided as chairman.[3] They met the candidates in early March, evaluating them on qualities such as academics, aptitude, pilot skills, character, and motivation. [4] The hopefuls had already undergone interviews in Houston, then a week of physical and psychological testing at Brooks Air Force Base in San Antonio. [5] In April, the selection board chose 19 new astronauts, including Ken Mattingly, Charlie Duke, Fred Haise, Stuart Roosa, Jack Swigert, Jim Irwin, Bruce McCandless, and Ed Mitchell.

Young and Collins continued their mission preparations until their complicated flight on July 18. The dynamics of the mission demanded a narrow launch window for Gemini X. The aggressive plan was for their docking target, Agena 10, to launch at 3:39 p.m. Houston time.[6] One hundred minutes later, Gemini X would launch and dock with the Agena

---

[3] Collins, *Carrying the Fire*, 178.
[4] Collins, *Carrying the Fire*, 178.
[5] Willie G. Moseley, *Smoke Jumper, Moon Pilot: The Remarkable Life of Apollo 14 Astronaut Stuart Roosa*, (Morley, MO: Acclaim Press, 2011), 92.
[6] Astronaut watches were synchronized to Houston time, even though they launched from the east coast.

on the fourth orbit at an altitude of about 160 nautical miles. It was no easy task since many Agena and Gemini launches had suffered significant delays, and only one successful docking had ever occurred in space.

If the docking went according to plan, Young and Collins would fire the Agena 10's engine to boost their orbit to over 400 nautical miles to take measurements in the lower portion of the Van Allen radiation belt. Then the Agena 10 engine would fire again to lower their orbit to about 200 nautical miles and rendezvous with Agena 8, the docking target from the Gemini VIII mission. Young would fly in formation while Collins performed a spacewalk over to Agena 8 and retrieve a package that had been recording micrometeorites in the four months since Neil Armstrong's docking. For all of those details to happen, Gemini X had to launch within a 35-second time frame, starting at 5:20:23 p.m. on July 18.

To the surprise of many, everything proceeded on schedule. Young and Collins were entering their spacecraft when they received word their Agena 10 launched and successfully achieved Earth orbit 160 nautical miles above. The crew waited patiently for their launch, hoping there would be no delays. Between the two launches, National Broadcasting Company (NBC) anchorman Frank McGee summarized the upcoming events:

> The Agena will be about 1,009 nautical miles ahead of the Gemini at the time of orbit insertion, and Agena Number 8 should be about 480 nautical miles behind it. And we'll have the three things whirling around up there together; hopefully all of them sufficiently synchronized so they can carry on the remainder of this mission. It is staggeringly complex.... They've done a lot of work getting ready for this one.[7]

NBC news correspondent David Brinkley continued, "It seems to me the age of computers had to arrive before the age of space. You certainly could not have figured all the variables with a paper and pencil, at least not fast enough." He later commented on the complicated series

---

[7] Lunarmodule5, "Gemini 10—Part 2 (NBC)," August 8, 2010, video, 9:56, https://www.youtube.com/watch?v=ck1n5q1r5Xw.

of events throughout the mission. "There is a great deal to be done. They have enough work to do the next three days to keep five or six people busy on the ground who have plenty of room to move around."[8]

The Gemini X liftoff occurred right on time and was a smooth experience. "There is no doubt about it, within one second of liftoff, you know you are moving," said Mike Collins in the post-mission debriefing. "There is a definite feeling of being booted from behind and a definite feeling of rising and a good feeling."[9] As the Gemini Titan raced toward orbit, it arced according to plan. Young and Collins were flying on their side, giving Collins in the right-hand Pilot seat a great look at Earth. "The boost part of the flight was a very pleasant surprise, and I enjoyed the heck out of it…. Just sitting up there watching the world go by, I thought that was great."

Young quickly added, "I told him to keep looking at those needles."[10]

As they entered Earth orbit, NASA's Public Affairs commentator Paul Haney gladly reported, "Mike Collins in jest suggested that we conclude this simulation and proceed with the debriefing," referring to the many hours of launch sims and debriefs preparing for all possible scenarios. "This, of course, is no simulation," Haney added.

Once in Earth orbit, there was no time to waste. Only a few minutes of daylight remained to check the status of all systems. As they moved into darkness, the crew began their first experiment—seeing if they could navigate to a target using their own star sightings without help from Houston.[11] Collins looked through a sextant, locating certain stars and measuring their angles to the horizon. These calculations would pinpoint the details of their orbit.[12] Collins strained as he attempted to distinguish the bottom of the atmosphere's airglow from Earth's terrain, and the process took longer than planned. The results were not as accurate as expected, so they used the Mission Control calculations instead.

---

[8] Lunarmodule5, "Gemini 10—Part 2 (NBC)."
[9] NASA, "Gemini X Technical Debriefing," (Houston: Manned Spacecraft Center, 1966), 3, https://digitalsc.lib.vt.edu/Ms1989-029/Ms1989-029_B05_F4a.
[10] NASA, "Gemini X Technical Debriefing," 8.
[11] Young and Hansen, *Forever Young*, 90.
[12] A sextant measures the angle between the horizon and celestial bodies for navigational purposes.

Two and a half hours after the launch, John Young adjusted their orbital plane to match that of Agena 10.[13] The crew spotted the docking target about 48 miles ahead. Young maneuvered the Gemini while Collins read him vital information such as distance from the Agena and their closing rate. Ideally, these two figures would reach zero simultaneously. Otherwise, they would miss the target. A slight error caused the Gemini to be out of plane with the Agena, forcing Young to make several adjustments within the last mile. Young said of the corrective procedures:

> I had to fly us through a *whifferdill*—a curlicue maneuver in which I would spiral Gemini X toward our target rather than taking a nice straight approach. Mike and I had done whifferdills in the simulator, and we knew it would take a darn good one to get us into the right position for our rendezvous. Worse yet, even a perfect whifferdill was a big waster of fuel. Because we were so out of plane, there was no other choice.[14]

The whifferdill was successful and their orbits were synchronized. Young nudged the nose of the Gemini into the collar of Agena 10, achieving the second dock in spaceflight history, joining Neil Armstrong from Gemini VIII in the elite club.

Young was frustrated by how much fuel the maneuver consumed. They planned on having 60 percent of their fuel left, but only 36 percent remained. He could not figure out why they needed those final corrections before docking.[15]

It was time to move on to the next item of business, testing the radiation levels at high altitude. Still docked, Young and Collins used the Agena engine to raise their orbit to the Van Allen Radiation Belt. Usually, astronauts do not see the rocket engine burn behind them. However, Young and Collins were nose-to-nose with the Agena, whose engine was on the far end of the stage. Young and Collins tightened their shoulder straps. The acceleration would be an "Eyeballs Out" sensation,

---

[13] Collins, *Carrying the Fire*, 206.
[14] Young and Hansen, *Forever Young*, 91.
[15] The answer came after the mission when a slight misalignment was discovered in the Gemini guidance platform, which contained a set of gyroscopes and fed information to the onboard computer assisting navigation.

pressing them forward against their harnesses compared to the normal "Eyeballs In" experience, which plastered them into their seats. They were flying over Hawaii when Collins entered the codes into the computer commanding the Agena to fire. The response was quite an experience. Young described the scene:

> It was right at sunset when we lit this baby off. There was a pop back there, then there was a big explosion, then a clang. We were thrown forward in the seats. We had our shoulder harnesses fastened. Fire and sparks started coming out the back end of that rascal. The light was something fierce, and the acceleration was pretty good. I never saw anything like that before, sparks and fire and smoke and lights. That is really a sight to behold.[16]

Collins was equally impressed with the Agena burn. "There is no subtlety to this engine, no gentleness in its approach. I am supposed to monitor the status display panel, but I cannot prevent my eyes from wandering past it to the glorious Fourth of July spectacle radiating out from the engine."[17]

The burn resulted in an elliptical orbit with a high point of 412 nautical miles, higher than any human before, reaching the Van Allen Radiation Belt. Young and Collins remained in that orbit overnight. They reported the first of many dosimeter readings measuring the effects of radiation, all of which were far lower than expected. This was good news for the upcoming Apollo flights to the Moon.

It had been a productive day, but the crew was discouraged as they prepared for the rest period. John Young kicked himself for using so much fuel during the rendezvous, while Mike Collins wondered if Houston would cancel one or both of the scheduled EVAs because of the lack of fuel remaining. All he could do was wait and see what Mission Control decided overnight.

The following morning, they woke up to good news: most major activities would proceed as planned. The second day was full of experiments. After breakfast and more negligible radiation readings, it

---

[16] NASA, "Gemini X Technical Debriefing," 47–49.
[17] Collins, *Carrying the Fire*, 212–213.

was time to reignite the Agena 10 engine for a braking maneuver lowering their orbit for a rendezvous with Agena 8 the following day. The engine bursting against their direction of travel was as spectacular as the day before. Collins said it "kicks like a mule," and Young added, "It may only be one G, but it's the biggest one G we ever saw. That thing really lights into you!"[18]

After the burn, Mike Collins received a "Go" to proceed with his first EVA. As they prepared, the astronauts spoke little on the radio, causing Director of Flight Crew Operations Deke Slayton to get on the loop: "You guys are doing a commendable job of maintaining radio silence. Since the French stopped shooting at you, why don't you do a little more talking from here on?"

After depressurizing the cabin in the dark, Collins stood up in the open hatch. "The stars are everywhere: above me on all sides, even below me somewhat, down there next to that obscure horizon. The stars are bright and steady," he said.[19] Using a 70mm camera with special high-speed film, he took a series of ultraviolet photographs of particular stars in the southern Milky Way. Each image was a 20-second exposure to study stellar ultraviolet radiation.[20]

After they passed from night to day, Collins photographed an eight-inch square titanium plate divided into four colors: red, yellow, blue, and gray. His pictures in direct sunlight would be compared to images shot of the same plate on the ground to evaluate film emulsions and developing techniques so future photographs in space would represent the actual colors.[21] After a few photos in the bright sunlight, Collins' eyes started to water. He mentioned it to Young, whose eyes were also watering. They decided to end the EVA a few minutes early and pressurize the cabin.

As they disconnected their oxygen hoses, they noticed the irritation had stopped. The ground surmised the problem was related to the crew running two compressors sending oxygen to their pressure suits. Houston recommended they only use one compressor during the next day's EVA.

---

[18] Collins, *Carrying the Fire*, 218.
[19] Collins, *Carrying the Fire*, 221.
[20] Hacker and Grimwood, *On the Shoulders of Titans*, 347.
[21] Collins, *Carrying the Fire*, 222.

They performed one more burn, circularizing their orbit to an altitude of roughly 205 nautical miles, just below the Agena 8 from the Gemini VIII mission. Since they were still docked to Agena 10 and able to use its engine, Young and Collins ended the day with 30 percent fuel remaining. Afterward, the tired crew ate dinner, performed housekeeping chores, and reviewed plans for the next day. They rested well during their second night in space.

Day three involved a rendezvous with the Agena 8 and Collins' second spacewalk. They used the Agena 10 engine at sunrise for a few small maneuvers to raise their orbit to intersect with their target, and then undocked. The Agena engine did not have the finesse needed for rendezvous, but they were grateful to use it and its fuel for a day longer than expected instead of the precious fuel onboard Gemini.

There were only 55 minutes of daylight in every orbit, so the timing for meeting up with Agena 8 was critical. The computations were made months before the flight. The goal was to see the Agena in the distance at dawn, draw near during the daylight pass, and arrive minutes before sunset. Collins calculated the needed corrections as he looked at the increasing size of the Agena through a sextant. He explains,

> By comparing its growth with my clock, I can give John a crude opinion of our range and closing velocity, but beyond that I can only shout encouragement and anti-whifferdill sentiments as we close in on it and he brakes to a halt. Finally, all motions are stilled, and we are riding serenely next to it, with 15 percent fuel remaining![22]

They arrived before the Sun slipped behind the horizon. John Young turned on the searchlight and kept the Agena in sight during darkness. Meanwhile, Mike Collins was busily preparing for his EVA which needed to begin at sunrise. He put on his EVA chest pack, a portable environmental control system regulating the oxygen in his suit. He attached it to a 50-foot umbilical cord containing an oxygen hose, a nitrogen hose, communication lines, and a tether. Right on schedule, they opened the hatch at dawn.

---

[22] Collins, *Carrying the Fire*, 227.

Once outside the spacecraft, the first item of business was for Collins to make his way by handrail to the rear of the Gemini to retrieve a micrometeorite detection plate that had been collecting data for two days. After handing it to Young, he headed to the back again to attach the nitrogen line to a handheld maneuvering unit, informally called a "zip gun," used for controlling his motions during the EVA. Shooting the pressurized gas in one direction would move him in the opposite direction.

The next step involved floating to Agena 8, about six feet away, to retrieve a micrometeorite package that had been exposed for four months. Collins carefully nudged himself away from the Gemini, landing by the Agena's nose. His momentum carried him off the nose, and he cartwheeled away from both vehicles. He used the zip gun to steady himself and maneuver back to the Gemini for another attempt.

Utilizing the maneuvering gun, he glided to the Agena and barely grasped it. He carefully made his way to the package again, being careful not to repeat the previous adventure. As he removed the package, the Agena began to tumble. Collins pulled himself back to the Gemini using the umbilical cord. As Mike stood in the hatch and collected his breath, John Young untangled the 50-foot umbilical cord. Then Collins tucked himself back in the Gemini, now filled with the umbilical. They stuffed the cord in a duffel bag along with the chest pack and other equipment no longer needed and discarded it through the hatch.

They slowly moved away from their target. Collins thought, "Adios, 8 Agena. We're finished with our three-month love affair with you, and now you are once again a derelict free to drift, a menace to be avoided."[23]

After the EVA, Young performed a burn to lower their orbit, preparing them for entry the next day. After completing a few more experiments, the astronauts prepared for their last night in space. Young and Collins had time to relax and enjoy their time in orbit for the first time since their busy flight began.

The final day went smoothly. The crew carefully reviewed their pre-retrofire checklist since everything had to be done exactly right. The retrofire burn slowed the spacecraft and brought it into Earth's

---

[23] Collins, *Carrying the Fire*, 236.

atmosphere. As the heat shield did its job, ionized gas surrounded the vehicle causing a communications blackout. The Gemini used its offset center of gravity to steer toward the landing point, rolling as directed by the computer.[24] At 38,000 feet, the six-foot diameter drogue parachute deployed, slowing and stabilizing the spacecraft. The 58-foot-diameter main chute deployed at 10,000 feet, lowering them at a rate of 30 feet per second. They landed three miles from their aiming point and within sight of the recovery ship *USS Guadalcanal.*

Gemini X accomplished many things which proved to be essential for the Apollo missions. Young and Collins practiced different types of rendezvous, successfully performed optical rendezvous with a passive target having no radar, and showed radiation levels were minimal on the outer edges of the Van Allen Belt. They also demonstrated one spacecraft could use the engine of another after docking and provided more valuable experience regarding the elusive EVAs.

Gemini XI flew two months later with Pete Conrad and Dick Gordon aboard. They achieved a rendezvous and docking with their Agena on the first orbit, only an hour and a half after liftoff. Gordon ran into problems on his first EVA as he attached a 100-foot tether between the Gemini and Agena. He tried to straddle the nose of the Agena while wedging his legs in the docking cone. That technique worked fine during zero-gravity training in an airplane flying parabolas, but it was a much bigger challenge in space. While trying to attach the tether, Gordon struggled to maintain his desired body position. Commander Pete Conrad could see his buddy was in trouble and ended the EVA early.

Without foot restraints and handholds, working outside the spacecraft caused extreme exhaustion. NASA needed to solve the problem before Buzz Aldrin's EVAs on the final Gemini flight. The crew trained in an underwater zero-gravity facility with a mockup of the mission hardware. Through tests, Gene Cernan verified it accurately reflected the problems he encountered on Gemini IX.[25] Extensive

---

[24] Martin Rush and Walter Vodges, "Three-Degree-of-Freedom Simulation of Gemini Reentry Guidance," *The Space Congress Proceedings 2*, 489, accessed July 1, 2020, https://commons.erau.edu/cgi/viewcontent.cgi?article=3224&context=space-congress-proceedings.

[25] David Schultz, Hilary Ray, Eugene Cernan, and Antoine Smith, "Body Positioning and Restrains During Extravehicular Activity," presented to the Gemini Summary

testing resulted in improved foot restraints and handrails, as well as waist tethers. The umbilical cord was shortened to 25 feet due to the problems with the bulky 50-footer. They also planned for rest periods during the EVA.

The result was three successful EVAs performed by Buzz Aldrin, including one lasting over two hours. In this EVA, the improvements allowed Aldrin to easily attach a tether between the Gemini and Agena target vehicle. Then Aldrin made his way to the back of the Gemini, attached his feet to the improved foot restraints, attached waist tethers, and completed several two-handed tasks such as tightening bolts and cutting cables. He moved forward to a workstation on the Agena and performed similar tasks, such as attaching connectors. He even wiped Jim Lovell's windshield. "Hey, would you change the oil too?" asked Lovell.[26] Aldrin demonstrated the capabilities of well-planned EVAs. The flight was the ideal ending to a successful Gemini program, and everyone entered the Apollo era with increased confidence and optimism.

Dr. Robert Gilruth, Director of the Manned Spacecraft Center, said after the flight, "We have done all the things we had to do as a prelude to Apollo. I believe the Gemini program has been most successful," adding Gemini passes along "a great legacy and birthright" to Apollo.[27]

Conference, February 1–2, 1967, 81, accessed July 1, 2020, https://ntrs.nasa.gov/api/citations/19680005472/downloads/19680005472.pdf.
[26] Gemini XII mission transcript, 44:42:33, https://historycollection.jsc.nasa.gov/JSCHistoryPortal/history/mission_trans/GT12_TEC.PDF.
[27] John Noble Wilford, "Last of Geminis Splashes Down Close to Target," *New York Times*, November 16, 1966, https://timesmachine.nytimes.com.

Illustration of the Gemini spacecraft approaching an Agena target vehicle. Artwork: Davis Paul Meltzer, NASA.

July 21, 1966: John Young (left) with Michael Collins aboard the recovery ship after the Gemini X mission. Photo credit: NASA S66-42777a.

Gemini X approaches the Agena 10 target vehicle 160 nautical miles above Earth. Photo credit: NASA S66-46122_G10-S.

Young and Collins had the rare opportunity to observe a burst from the Agena engine. Photo credit: NASA S66-46249_G10-S.

5

NEW ASTRONAUTS

THE ORIGINAL NINETEEN

On May 2, 1966, while John Young and Mike Collins were preparing for their Gemini X mission, the new 19 astronauts reported for duty. Several of the Original 7 Mercury astronauts greeted them and their families. It left quite an impression on Charlie Duke and his wife, Dotty. Duke said,

> We were just agog. Here was Alan Shepard—the first American in space—an authentic American hero! We were speechless, and Dotty's eyes got real big. I was really impressed. *There was Deke Slayton, Gordon Cooper, Wally Schirra. Wow, and now I'm one of them!* When Dotty and I got back into the car, she excitedly exclaimed, "Imagine, meeting Alan Shepard—a real astronaut!" I reminded her I was an astronaut too, but I knew what she meant.[1]

Charlie Duke was raised in Lancaster, South Carolina, a textile town at the time. "Most people did not go to college, much less go to the Moon. So I did not go out in the backyard and say, 'Mama, one day I'm

---

[1] Charlie Duke and Dotty Duke, *MoonWalker: The True Story of An Astronaut Who Found That the Moon Wasn't High Enough to Satisfy His Desire for Success*, (Nashville: Thomas Nelson, 1990), 89. Emphasis his.

going to land on the Moon.' Mama would have sent me to the psychiatric hospital!"[2]

Following the attack on Pearl Harbor in December 1941, Charlie's dad volunteered for the Navy at the age of 33. Charlie was extremely patriotic as a young boy, and that sentiment grew through his involvement with the Boy Scouts. "Scouts were a big part of my upbringing. I had a great Scoutmaster, Mr. Bundy. He and other leaders were big character builders," said Duke. He learned a lot by being outside on frequent camping trips, even spending the night alone in the woods. He achieved the highest rank of Eagle Scout at the age of 14.[3]

"My dad was a good guy, and he encouraged my brother and me, but he never got involved," said Duke. "He didn't go camping with us, didn't go hunting with us. He was more of a businessman, plus he loved to cook. His weekends were spent cooking, not exotic dishes, but very complicated ones. He even made his own spaghetti sauce, so it was an all-day process." There was a lot of land near their house, so Charlie's dad bought a 20-gauge double-barrel shotgun for him and his identical twin brother Bill. "We would load up with some shells and take turns shooting rabbits and quail, and that prompted my love for hunting which carried on all my life."[4]

He decided to attend the Naval Academy, and after receiving an appointment from a local Congressman, Duke went to a prep school in Florida to help him prepare. After graduating as valedictorian from Admiral Farragut Academy in St. Petersburg, he began his studies at the Naval Academy. He loved his time there. It was an excellent education and a significant time of growing up.

Duke had sea duties each summer and realized the cruises made him sick. Halfway through his time at Annapolis, he was given a ride in a bi-wing open-cockpit seaplane. Soaring over the Chesapeake Bay was a life-changing experience for him, and he began to consider aviation as a career.

By his senior year, he had fallen in love with airplanes and knew he wanted to be a pilot. The only question was whether he would serve

---

[2] Charlie Duke, Address to Southside Community Church, Corpus Christi, Texas, September 8, 2018.
[3] Charlie Duke, interview by author, May 20, 2020.
[4] Charlie Duke, interview by author, May 20, 2020.

in the Navy or the Air Force. Duke took a physical examination with the Navy. They detected astigmatism in his right eye and informed him, "You don't qualify for Naval Aviation, but the Air Force will take you." He has had numerous physical exams over the years, especially at NASA, and nobody has detected astigmatism since then.[5]

After graduating with distinction from the Naval Academy in 1957, Charlie Duke was commissioned into the United States Air Force. He excelled in flight training. As one of the top students in his class, he was rewarded with the assignment to be a fighter pilot. Flying came naturally for Duke, and he loved his three-year deployment to Germany. While a fighter interceptor pilot at Ramstein Air Base in Germany, he read about the US space program. In May of 1961, President Kennedy gave the challenge to land on the Moon by the end of the decade. Duke said, "Everyone in my squadron laughed at him. We only had 15 minutes in space, and he's sending us to the Moon."[6]

At that time, the Air Force recommended their young officers pursue advanced degrees. Duke moved to Cambridge, Massachusetts, where he began graduate school at the Massachusetts Institute of Technology, MIT. He earned a Master's degree in Aeronautics and Astronautics, then was accepted as a test pilot at the Aerospace Research Pilot School at Edwards Air Force Base, where Chuck Yeager was the Commandant. After graduating in 1965, Duke became a flight instructor at Edwards, where he was serving when selected by NASA.

Despite his impressive list of achievements, perhaps Duke's greatest success was winning the heart of Dorothy Claiborne, whom he met in Boston. Charlie and Dotty were married on June 1, 1963, at St. Phillip's Cathedral on Peachtree Street in her hometown of Atlanta, Georgia. Duke suffered from food poisoning on their honeymoon in Jamaica. "At first you think you are going to die, but then things get so bad you want to die but are afraid you won't," he wrote in his autobiography.[7]

It was a preview of a challenging first year of marriage, full of medical catastrophes and long hours with the demanding workload of MIT graduate studies. Early in their marriage, it was clear they had two

---

[5] Charlie Duke, Southside Community Church, September 8, 2018.
[6] Charlie Duke, Southside Community Church, September 6, 2018.
[7] Duke, *MoonWalker*, 63.

different goals. Dotty pursued their relationship while Charlie pursued his career. Like many in the NASA community, their marriage would be severely tested in the years to come.

Also in the group of 19 new astronauts was Ken Mattingly. If there was ever a young boy destined to be a pilot, it was Mattingly. "As a kid, my earliest memories that I can recall all had to do with airplanes. My dad worked for Eastern Airlines. Before I had any idea what that was, my toys were all some kind of airplane, and any picture that you could glean from when I was a child, they always had an airplane in it."[8]

He spent his free time going to the airport and watching the planes. He built every model airplane he could get his hands on. "If I could stay out of trouble for a while, my reward was Dad had passes on Eastern, and I would get on the airplane, and I would just fly to the end of the route and back. Never got off the airplane, just go from one place to the other."[9]

He attended Auburn University on a Navy ROTC scholarship, majoring in aeronautical engineering. After graduation, he entered the Navy as a commissioned officer. He was devastated when not selected for flight training. Instead, he was sent to the *USS Galveston,* working in the gunnery department with sophisticated weaponry. One day Mattingly's supervisor pulled him aside and asked, "You know, I understand you don't do anything but read airplane books. Why don't you go to flight training?" When Mattingly told him he applied but was not chosen, the boss checked into it and discovered there was a mix-up. He told Mattingly to take the next day off and clear it up.

Mattingly needed everyone up the chain of command to sign off on the change. The final stop was the gunnery officer, Lieutenant Commander Glenwood Clark. Clark was convinced airplanes were no longer relevant, and the future of naval warfare rested in the gunnery system on which Ken had been working. Mattingly politely agreed but insisted he still wanted to go to flight training. Glenwood Clark told him, "You are the dumbest ensign I have ever met." He signed the necessary paperwork and sent Mattingly on his way.[10]

[8] Thomas K. Mattingly, II, interview by Rebecca Wright for NASA Johnson Space Center Oral History Project, November 6, 2001, 1.
[9] Ken Mattingly, interview by Rebecca Wright, November 6, 2001, 2.
[10] Ken Mattingly, interview by Rebecca Wright, November 6, 2001, 4.

Flight training went well. A few years later, he was stationed at Naval Air Station Jacksonville. A friend in a photo-reconnaissance squadron went to Cape Kennedy to photograph the Gemini III launch and asked Mattingly to join him. Mattingly listened to the air-to-ground conversation while watching the spacecraft leave the F-4 chase planes in the dust. He thought, "That sounds like the most exciting thing anybody could ever do."

Mattingly eventually resigned from the Navy and attended the Air Force Test Pilot School at Edwards Air Force Base, where Hank Hartsfield and Charlie Duke were instructors. "There was nothing else to do out there in the desert except fly, so it was really, really a lot of fun."[11] During his time at Edwards, NASA announced they would be selecting another class of astronauts. Mattingly applied along with several others from Edwards. After physical exams and some classwork, he was invited to Houston for the interview phase.

"That was a fascinating little experience, my first introduction to John Young and Mike Collins. I didn't figure John out that day and didn't do so for a long time after that, either. I figured I was dead meat."[12] Fortunately, the selection board did not agree.

A few days later, Mattingly received a call from Deke Slayton with the great news NASA had selected him. Slayton told him not to tell anyone until it was announced publicly, but Mattingly's friends knew immediately from the large grin on his face. They celebrated at a restaurant on the Riverwalk in San Antonio, Texas. Mattingly said,

> We're sitting around this table, and the waiter comes over. He handed me a piece of paper. He said, "This lady over here would like your autograph." Somebody recognized me? Wait a minute. How does she know? I said, "Who does she think I am?" He said, "Aren't you Dickey Smothers?" I said, "No, but let me give you this and just hang onto it for a week."[13]

After arriving at NASA, Ken Mattingly, Charlie Duke, Stu Roosa, and the other new astronauts started four months of academic courses, including spacecraft systems and orbital mechanics. They also began

---

[11] Ken Mattingly, interview by Rebecca Wright, November 6, 2001, 8.
[12] Ken Mattingly, interview by Rebecca Wright, November 6, 2001, 9.
[13] Ken Mattingly, interview by Rebecca Wright, November 6, 2001, 11.

their study of geology. The astronauts supplemented their geology coursework with field trips to the Grand Canyon, Oregon, and Alaska during the summer of 1966.

On October 3, 1966, the new astronauts received their engineering oversight assignments. Mattingly, fascinated by Ed White's spacewalk on Gemini IV, was excited to help with the space suit development. He said, "It started one of the most fascinating and rewarding periods that I've had."[14]

Charlie Duke and Stu Roosa were assigned to work under Frank Borman overseeing the progress of the Saturn V launch vehicle. They made frequent trips to Huntsville, Alabama, to observe the work done by Wernher von Braun and his team. Charlie Duke assisted at the Booster console for the launch of the last two Gemini flights in September and November 1966, while Roosa served as a CAPCOM for Gemini XII.

As the first Apollo flights neared, significant issues still plagued the spacecraft. The Lunar Module (LM) development was far behind schedule and was the "pacing item," holding up the entire program. The Apollo Command Module (CM) had its share of problems. Unlike the Gemini capsules, which left the contractor plants tested and almost ready for flight,[15] the more complex Command Module needed plenty of engineering work once received by NASA.[16]

The most troublesome aspect of the spacecraft was the Environmental Control Unit (ECU), the heart of the Environmental Control System. The system provided a pressurized cabin for the crew, a 100 percent oxygen atmosphere, comfortable temperature in the cabin, hot and cold water, carbon dioxide removal, and the venting of waste. It also prevented the electronic equipment from becoming overheated.[17]

---

[14] Ken Mattingly, interview by Rebecca Wright, November 6, 2001, 14.

[15] Hacker and Grimwood, *On the Shoulders of Titans*, 384.

[16] Courtney G. Brooks, James M. Grimwood, and Lloyd S. Swenson, Jr., *Chariots for Apollo: A History of Manned Lunar Spacecraft*, (Washington, DC: National Aeronautics and Space Administration, 1979), 209.

[17] NASA, *Apollo Spacecraft News Reference: Command and Service Modules*, Prepared by the Space Division of North American Rockwell Corporation, Downey, CA, in cooperation with NASA's Manned Spacecraft Center, (Los Angeles: Periscope Film, LLC, 2011), 117.

The ECU was an enormously complicated design. The unit was 29 inches long, 16 inches high, 33 inches wide, and weighed 158 pounds. Packed inside were a water cooler, water-glycol evaporators, lithium hydroxide canisters, and a suit heat exchanger.[18] The ECU on the Apollo 1 Command Module, serial number 012, had issues and was replaced before the spacecraft was shipped to NASA on August 25, 1966.[19]

NASA Astronaut Group 5, selected in April 1966, "The Original Nineteen." Front row, from left to right: Edward Givens, Ed Mitchell, Charlie Duke, Don Lind, Fred Haise, Joe Engle, Vance Brand, John Bull, Bruce McCandless. Back row left to right: Jack Swigert, William Pogue, Ron Evans, Paul Weitz, Jim Irwin, Gerald Carr, Stu Roosa, Al Worden, Ken Mattingly, Jack Lousma. Photo credit: NASA.

North American Aviation technicians prepare to install the crew compartment heat shield on the Apollo 1 Command Module. Photo credit: NASA S66-41851, Kipp Teague, scan by Ed Hengeveld.

[18] NASA, *Apollo Spacecraft News Reference: Command and Service Modules*, 119.
[19] Brooks, Grimwood, Swenson, Jr., *Chariots for Apollo*, 209.

Ken Mattingly (left) with John Young.
Photo credit: NASA S72-19418, Kipp Teague, scan by J.L. Pickering.

Apollo 1 prime crew, left to right: Gus Grissom, Ed White, Roger Chaffee.
Photo credit: NASA AP1-S66-24522, Kipp Teague.

6

SHAKEN TO THE CORE

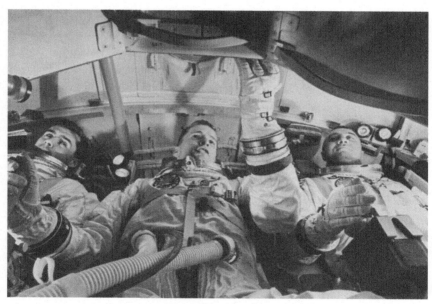

Apollo 1 astronauts Roger Chaffee, Ed White, and Gus Grissom.
Photo credit: Life Magazine.

JANUARY 27, 1967

Gus Grissom commanded the Apollo 1 crew, joined by Ed White and
rookie Roger Chaffee. They would fly a simplified Command Module,
called a Block I, designed for Earth orbit only and lacking any docking
mechanism. NASA planned two Block I flights. Then the Block II
Command Module, equipped for the more complicated lunar missions,

would be introduced. Grissom's crew had been testing their CM number 012 since it was on the production line at the North American plant in Downey, California. They were concerned about the condition of the spacecraft, especially the wiring. So was John Young.

On Friday, January 27, 1967, Young was in Downey evaluating the other Block I Command Module, scheduled for the second manned Apollo flight. He was on the backup crew for Apollo 2, along with Tom Stafford and Gene Cernan. They ran tests on the CM, taking turns with the Apollo 1 backup crew of Jim McDivitt, Dave Scott, and Rusty Schweickart. In the middle of a test run, Young noticed some drops of glycol dripping from the Environmental Control System and puddling onto the floor. They were discouraged but pressed on.[1]

The same day, Apollo 1 prime crew members Grissom, White, and Chaffee were at the Cape in their spacecraft, 012. They were performing a "plugs out" test, where external power supplies were removed and the spacecraft would use its internal power. It was a dress rehearsal for the countdown of the actual flight scheduled for February 21. The last part of the test would be a simulated emergency to determine how quickly the astronauts could escape from the Command Module. Pyrotechnics expert Sam Beddingfield would oversee that portion.[2]

Years before, Beddingfield was an Air Force test pilot at Wright–Patterson Air Force base in Ohio with Gus Grissom. Grissom was selected as a Mercury 7 astronaut and encouraged Beddingfield to work at the space agency with him. Beddingfield was hired by NASA in 1959 and served as an engineer during the Mercury program. When Grissom's hatch blew at the end of the second Mercury flight, Beddingfield defended Grissom's account of the hatch exploding on its own. As one of the official investigators, Beddingfield identified two series of events verifying Grissom's testimony that it blew without hitting the plunger.[3] Wally Schirra's injured hand after hitting his own plunger was proof of Grissom's innocence.

Beddingfield was among those monitoring the Apollo 1 tests from the blockhouse near Pad 34. Problems delayed the countdown, so he slipped away for dinner at a restaurant in Cocoa Beach. Returning down

---

[1] Young and Hansen, *Forever Young*, 111.
[2] Murray and Cox, *Apollo*, 222.
[3] Murray and Cox, *Apollo*, 222.

the narrow road to Pad 34, Beddingfield was overtaken by an ambulance racing toward the launch tower. He arrived to chaos in the blockhouse. A fire had erupted in the spacecraft.

At 6:31 p.m. local time, while the crew was troubleshooting communication problems, a spark ignited combustible material in the spacecraft. The cabin contained 100 percent oxygen pumped up to over 16 psi to avoid damage to the CM from the sea-level atmosphere.[4] Flames quickly erupted.

In the blockhouse, Deke Slayton was startled by the word "Fire," then Roger Chaffee's voice, "We've got a fire in the cockpit!" Monitors showed the CM hatch window with Ed White reaching for the inner hatch handle. Slayton's nightmare intensified as he heard, "We've got a bad fire... we're burning up!"

Stu Roosa was CAPCOM in the blockhouse and sat beside Slayton. Roosa tried calling the crew but received no answer. Chris Kraft was at the Flight Director console in Houston, listening on three separate communication loops. "I heard a voice yelling, 'Get us out of here!' I felt sure that it was Roger," recalled Kraft. "The sounds of confused scuffling and movement echoed in my headset, along with reports of a fire in the 'White Room'—the clean room surrounding the spacecraft. I started praying, harder than I'd ever prayed in my life."[5]

The technicians engaged in heroic measures attempting to open the hatch. Kraft recognized the voice of test conductor George Page on the loop. "I heard his words, and I heard despair." Kraft asked him about the crew. Page responded, "Not much hope, Flight. We'll have the hatch open in a minute." Soon afterward, Kraft's worst fears were realized. "The crew is dead."[6]

Christopher Kraft was shaken to his core. "We'd put three astronauts into harm's way and made their escape impossible. They were dead, and we knew it was our fault. The fire on the pad and its consequences would be profound."[7]

---

[4] Christopher C. Kraft, *Flight: My Life in Mission Control*, (New York: Dutton, 2001), 273.
[5] Kraft, *Flight*, 270.
[6] Kraft, *Flight*, 270.
[7] Kraft, *Flight*, 270.

Slayton and Roosa hurried to the launch pad, and the scene would be carved into their memories for the rest of their lives. Roosa noticed the spacesuits were still white. The astronauts had sustained burns, but they were not life-threatening. They died from inhaling carbon monoxide from the fire. The positions of the three men showed they followed their training for emergency egress, with Chaffee still buckled in his seat, staying in communication with CAPCOM. White and Grissom were trying to open the hatch. Roosa believed the media's reporting of charred bodies amounted to journalistic irresponsibility, bringing extra pain to the families.[8]

In Houston, Mike Collins was part of a meeting in Deke Slayton's office when the red phone rang. Slayton's assistant, Don Gregory, answered, listened quietly, then stated, "Fire in the spacecraft." Nothing more needed to be said.

Collins and fellow astronaut Alan Bean arranged the notification of the wives. Collins lived three doors from his close friend Roger Chaffee and wife, Martha. Collins knew he should be the one to tell Martha. He drove slowly to their home. Sue Bean and other wives from the neighborhood were there for support when Collins arrived. "Martha, I'd like to talk to you alone...."[9]

The bodies were removed sometime after 1:00 a.m. and taken to a makeshift morgue a mile away. Slayton flew back to Houston with Roosa. Deke was close with all the crew members, but especially Grissom. "It was a bad day," said Slayton. "Worst I ever had."[10]

True leadership shines in the crucible of adversity. This was certainly the case with NASA. On the Monday morning after the fire, Flight Director Gene Kranz called a meeting with the flight controllers. His message was memorable:

> Let us get good and angry—and then let us make no mistakes.... From this day forward, Flight Control will be known by two words: *Tough* and *Competent*. *Tough* means we are forever accountable for what we do or what we fail to do. We will never again compromise our responsibilities. Every

---

[8] Moseley, *Smoke Jumper*, 103–104.
[9] Collins, *Carrying the Fire*, 270–271.
[10] Donald K. Slayton and Michael Cassutt, *Deke! US Manned Space: From Mercury to The Shuttle*, (New York: Tom Doherty Associates, 1994), 190.

time we walk into Mission Control, we will know what we stand for. *Competent* means we will never take anything for granted. We will never be found short in our knowledge and in our skills. Mission Control will be perfect.

When you leave this meeting today, you will go into your office and the first thing you will do there is to write *Tough* and *Competent* on your blackboards. It will *never* be erased. Each day when you enter the room, these words will remind you of the price paid by Grissom, White, and Chaffee. These words are the price of admission to the ranks of Mission Control.[11]

Roger Chaffee and Gus Grissom were interred at Arlington National Cemetery on January 31. Ed White, a classmate of Mike Collins at the US Military Academy, was laid to rest the same day at West Point Cemetery. All were buried with full military honors.

REVIEW BOARD

NASA immediately formed a Review Board, with Frank Borman as the astronaut representative, to investigate the cause of the accident and generate recommendations for making the spacecraft as safe as possible. On April 5, they submitted their three-thousand-page report, finding:

* The spark igniting the fire likely occurred on the lower left side of the spacecraft "where Environmental Control System (ECS) instrumentation power wiring leads into the area between the Environmental Control Unit (ECU) and the oxygen panel."
* The official cause of death was asphyxia due to breathing toxic fumes from the fire.
* Due to the rising pressure caused by the fire, the astronauts could not budge the inward-opening hatch.
* Procedures had not been developed for this type of emergency.[12]

---

[11] Kranz, *Failure Is Not an Option*, 204.
[12] NASA, "Report of the Apollo 204 Review Board to The Administrator, National Aeronautics and Space Administration," accessed August 5, 2019, https://www.history.nasa.gov/Apollo204/summary.pdf, 6-1.

The Review Board noted several issues with the ECS. "Components of the Environmental Control System installed in Command Module 012 had a history of many removals and technical difficulties." They also mentioned, "Coolant leakage at solder joints has been a chronic problem" and, "The coolant is both corrosive and combustible."[13] The Review Board had numerous recommendations, including limiting combustible material in the Command Module, replacing the 100 percent oxygen with a mixture of nitrogen and oxygen for tests on the ground, developing procedures for such an emergency, and changing to a simplified outward opening hatch.[14]

The hatch in the Apollo 1 spacecraft was bulky two-piece construction. The interior part opened inwardly, requiring the insertion and cranking of a ratchet handle in six locations. In the best conditions, it took more than a minute to open. The designers had been more concerned about preventing the hatch from accidentally opening in space or after landing in water than they were about a quick escape on the ground.

The fire caused everyone to look at the spacecraft in a new way, and a safer Command Module emerged. Later, Gene Kranz would write,

> As we fought back from the tragedy, *Tough* and *Competent* joined with *Discipline* and *Morale* in defining the culture of the controllers. These words became our rallying cry.... The ultimate success of Apollo was made possible by the sacrifices of Grissom, White, and Chaffee. The accident profoundly affected everyone in the program. There was an unspoken promise on everyone's part to the three astronauts that their deaths would not be in vain.[15]

GEORGE LOW

The fire exposed severe flaws in the Command Module design, manufacture, and quality control. A complete overhaul was necessary. In April 1967, George Low was appointed Manager of the Apollo Spacecraft Program Office (ASPO), where he would address the issues

---

[13] NASA, "Report of the Apollo 204 Review Board," 6-3.
[14] NASA, "Report of the Apollo 204 Review Board," 6-1, 6-2, 6-3.
[15] Kranz, *Failure Is Not an Option*, 205.

with the Command Module and Lunar Module, which was far behind schedule.

George Low's importance to the Apollo program's success cannot be overemphasized. His biographer Richard Jurek describes Low as "one of the indispensable people behind the success of Apollo, especially after the Apollo 1 fire. It is no wonder that Frank Borman has pointed to Low as the one human most responsible for Apollo's success or that Neil Armstrong called him his favorite engineer."[16] Gene Kranz described Low as "a master of getting people to work together, creatively channeling their energies and thus building the momentum to achieve the objective."[17]

Low became NASA's Chief of Manned Spaceflight in 1958. He was heavily involved in setting long-range goals for Projects Mercury, Gemini, and Apollo and was one of the earliest advocates for lunar landing missions. "Low felt that a Moon landing would provide the stretch goal that would be not only worthy of the financial costs but also worthy of the potential human risk and cost," said Jurek.[18] When many questioned if NASA should continue manned space flights after Project Mercury, Low chaired a committee recommending a lunar landing. Their feasibility study guided President Kennedy to make the end-of-decade challenge.[19] The fire put the goal at risk.

The tragedy sunk the development of the Apollo spacecraft into disarray. NASA needed Low's skill, drive, and experience to right the ship. Richard Jurek explains:

> The agency was mired in anger, doubt, and paralysis. Low's calm, quiet, yet confident leadership helped to cut through the competing and complex challenges and laser-focused the team on the task at hand. And he did so with an unprecedented attention to detail and personal responsibility that he helped to instill across the entire supply chain, from contractor to NASA official. He was the right person, at the right time when the

---

[16] Richard Jurek, email message to author, February 23, 2021.
[17] Kranz, *Failure Is Not an Option*, 250.
[18] Richard Jurek, email message to author, February 23, 2021.
[19] Jurek, *The Ultimate Engineer*, 78–79.

program needed a miracle worker of his unique skill set and ability.[20]

Low knew time was running out if they were to meet President Kennedy's deadline. He worked eighteen hours a day, seven days a week, to improve the program office communication with the engineering and operation teams and reign in the number of design changes that hampered the spacecraft's development. He tightened up the relationship with contractors, increasing communication and accountability. Low's dealings with contractors were more contractual than in the past. Low also improved the weekly Configuration Control Board (CCB) meetings. His CCB consisted of top NASA leadership, including Chris Kraft (Director of Flight Operations), Max Faget (chief engineer in Houston), Scott Simpkinson (Flight Safety), Deke Slayton (head of the astronauts), Low's two deputies Ken Kleinknecht and Rip Bolender, plus Dale Myers and Joe Gavin. They were program managers from North American and Grumman, manufacturers of the Command Module and Lunar Module.

Low's Configuration Control Board met every Friday at noon from June 1967 until July 1969. The heated meetings often lasted into the night but effectively reduced the constant revisions. Low explained the process:

> We dealt with changes large and small, discussed them in every technical detail, and reviewed their cost and schedule impact. Was the change really necessary? What were its effects on other parts of the machine, on computer programs, on the astronauts, and on the ground tracking systems? Was it worth the cost, how long would it take, and how much would it weigh?"[21]

As NASA adapted to challenges, a "Lead Center Concept" emerged where the Apollo Spacecraft Program Office CCB in Houston dealt not only with spacecraft (CSM and LM), but also other elements of the Apollo program. Sam Phillips and others wanted the change control decentralized away from Washington so key decisions would be

---

[20] Richard Jurek, email message to author, February 23, 2021.
[21] Cortright, *Apollo Expeditions to the Moon*, 73.

made at the field centers such as Houston, Huntsville, and the Cape, where the experts were located. "There were project program offices at Marshall [Huntsville] for the launch vehicle and also at Kennedy for the launch facilities, but they all reported to George Low at the center, not headquarters," said Gerry Griffin.[22] George Low and his CCB were uniquely qualified to oversee the operation. "You had to have the right kind of person at the field center to do that, and we were lucky we had George. He was extremely effective because everyone respected him," Griffin added.[23]

Ken Mattingly says, "George Low is the finest program manager that ever walked the face of the Earth. I've seen a lot of good people, but I've never seen anybody of his caliber." Mattingly continues, "Low had this way that he could take that information and use it, but never embarrass anybody or never expose. He would just ask questions based on his knowledge that there was more to this story than perhaps was immediately obvious."[24]

Various projects were spread out to field centers and contractors across the country. Change control started at the bottom and worked its way up when necessary. Rules clearly defined what each level could approve without going up to the next level, and the goal was to keep the decisions as low as possible. The ground-level change control happened with the contractors. As the pieces were assembled into subsystems and systems, their CCBs were located in the corresponding field centers. George Low's top-level CCB in Houston dealt with major projects like the Lunar Module, Command Module, Saturn stages, and issues affecting launch dates.[25]

DOWNEY, CALIFORNIA

On April 27, 1967, Frank Borman was appointed to lead a "redefinition team" overseeing the renovations on the Command Module at the North American plant in Downey, California. Borman had spent months with

---

[22] Gerry Griffin, interview by author, August 9, 2020.

[23] Gerry Griffin, interview by author, August 9, 2020.

[24] Mattingly, Thomas Kenneth, II, interview by Kevin M. Rusnak for NASA Johnson Space Center Oral History Project, April 22, 2002, 2.

[25] Gerry Griffin, interview by author, August 9, 2020;
NASA, "Apollo Configuration Management Manual," 5.1.3.1.

the Review Board after the fire; he knew precisely what was needed to fix the spacecraft. He and his team relocated to Downey for the better part of a year implementing the recommendations of the Review Board.

Significant changes to the spacecraft included a unified hatch designed to open outwardly in less than 10 seconds. They greatly improved the routing and quality of the wiring, and combustible materials in the spacecraft and space suits were replaced by non-flammables. Other important advancements included a change from 100 percent oxygen on the launch pad to a less-volatile 60 percent oxygen and 40 percent nitrogen mixture. A system gradually replaced the cabin environment with 100 percent oxygen after launch.[26]

"I think I contributed the most to the Apollo Program out in Downey, California, as part of the redefinition team, not as an astronaut on Apollo 8," Borman remembers. "The time at Downey was the most challenging. We had a lot of balls in the air. Our team out there did a really important job."[27]

While Borman's team was going through every aspect of the Command Module, flight controllers saw an opportunity to address one of their concerns with the spacecraft—insufficient instrumentation. "In Mission Control, we had noted for a long time that we were short on instrumentation in some areas," said Gerry Griffin, a Guidance, Navigation, and Control (GNC) Officer. "We thought they could add some temperature and pressure sensors without having to tear up too much to give us a little more insight into some of the Environmental Control System and some on the propulsion side."[28]

Griffin was designated to go to Downey and work not on Borman's team but adjacent to it. "I was focused on the instrumentation improvements while they were inside the guts of the spacecraft. We tried to do all of that without tearing up anything that wasn't going to be torn up in order to get these other fixes in."[29] He added, "The only instrumentation changes we were able to get approved could be called

---

[26] Frank Borman, *Countdown: An Autobiography*, with Robert J. Serling, (New York: Silver Arrow, 1988), 185–186.

[27] Frank F. Borman, interview by Catherine Harwood for NASA Johnson Space Center Oral History Project, April 13, 1999, 56.

[28] Gerry Griffin, interview by author, July 13, 2020.

[29] Gerry Griffin, interview by author, July 13, 2020.

'targets of opportunity'—simple changes requiring little additional work, little or no cost, and no impact to the schedule."[30]

Gerry Griffin estimates he flew to California at least 30 times for his work at the Downey plant. In those days, Continental Airlines had nonstop flights from Houston to Los Angeles, and their marketing slogan was "The Proud Bird with the Golden Tail." Since Griffin was often out of town flying Continental, the other flight controllers nicknamed him "The Golden One."

A few months later, Griffin was selected as a Flight Director and had to choose a color designation for his team. He intended to choose Maroon in honor of his beloved alma mater Texas A&M University, but the flight controllers said, "Gerry, you've got to pick Gold."[31] Choosing the camaraderie within Mission Control over his school loyalty, Griffin decided to become Gold Flight Director.

While manned flights were on hold, NASA took advantage of the time to get other matters in order, such as planning the sequence of upcoming Apollo flights, training astronauts, implementing tighter management controls, and regaining the momentum from the previous year.

NASA formed the Emergency Egress Working Group, a special committee to review launch emergency procedures. Charlie Duke and Stu Roosa were the astronaut representatives on this project. Duke said, "Our task was to help redefine and develop procedures for escape on the pad so this kind of tragedy wouldn't happen again. Over the next year we were very active, attending meetings weekly and developing new escape techniques and equipment."[32]

The group developed two ways off the spacecraft level of the launch pad in case of an emergency. One was a slide-wire connected to the launch tower, where the astronauts and technicians would run across the swing arm and attach themselves to the slide-wire system. Duke describes early versions as "everybody grabbed a horse collar, put it over you, hooked it onto the slide wire, and jumped off."[33] The final version

[30] Gerry Griffin, email message to author, February 26, 2021.
[31] Gerry Griffin, interview by author, July 13, 2020. Griffin's choice left "Maroon" open for new Flight Director Milt Windler, who graduated from Virginia Tech University, another school with maroon colors.
[32] Duke, *MoonWalker*, 97.
[33] Charlie Duke, interview by author, May 20, 2020.

had a nine-person basket, or gondola, attached to the wire. Duke recalls, "The big joke at the time was, 'What if the first guy in the basket cuts it loose and the other guys are trapped up there?'"

The second way of escape was the elevator, which would descend at high speed to the base of the tower. Then a slide would take them down under the pad to a large blast room suspended on shock-absorbing springs, sealed off by a massive door. "It was like being in a bank vault, but it had enough water, oxygen, and food to last 30 days," said Duke. "If there was a full explosion above you, the springs in this room would dampen the shock, and if all that debris fell on top of the blast room, it might take them a month to dig you out. It was a pretty impressive room."[34]

LUNAR MODULE

Several important events in August 1967 enhanced the odds of meeting President Kennedy's deadline. One area of improvement involved the Lunar Module, which had fallen behind schedule. The bottom half of the complicated LM was the descent stage containing four spindly legs and an engine with a throttle that would adjust power as it gently lowered the spacecraft to the lunar surface. When it was time for lunar liftoff, the descent stage would remain on the Moon and serve as a launch pad for the ascent stage, which contained its own engine and the crew cockpit.

The ascent engine was the most critical in the Apollo program. There was no backup mode of operation in case it failed to lift the crew from the lunar surface. It had to work, or else the crew would be stranded. Bell Aerospace Corporation built the engine but could not solve the problem of combustion instability. The issues centered around the injector, which forced the propellants into the combustion chamber.

On August 1, 1967, Low made public his decision that the Rocketdyne Division of North American Aviation would design and develop an alternate injector for the Bell engine. This parallel program would give NASA two injector options in case Bell could not get the problem under control. Low assigned a team to oversee the progress of the ascent engine injector. Since Charlie Duke had spent a lot of time

---

[34] Charlie Duke, interview by author, May 20, 2020.

working with the Saturn propulsion systems, he was a natural choice to be the astronaut's representative on Low's committee. Over the next nine months, they traveled often to the Bell plant in New York and Rocketdyne in California to assess the situation.

On August 3, Low made another giant stride toward the Moon by announcing Bill Tindall's appointment as "Chief of Apollo Data Priority Coordination," commonly known as Mission Techniques. Tindall's role was to figure out the details of flying a lunar mission by coordinating input from Low's Apollo Spacecraft Program Office, the Flight Control Operations (directed by Chris Kraft), and Flight Crew Operations (directed by Deke Slayton).[35]

Tindall was an expert in orbital mechanics and became an integral part of Apollo's success. Gene Kranz had a great appreciation for him. Flight controller Steve Bales recalls, "If there were ever two people who were different in personality, it was Gene Kranz and Bill Tindall. Gene is military from head to toe, and Bill is a people person from head to toe. They were really different, but they became very close friends."[36] Kranz fully understood how important Tindall was to Apollo. On July 20, 1969, Gene made sure Bill sat beside him as Kranz directed the first lunar landing, a gesture that meant the world to Tindall.

Tindall had a magical way of synthesizing diverse views from opinionated engineers, flight controllers, and astronauts. He would assemble them in one room, throw out scenarios, and let them have at it. For example, "We've just arrived at the Moon, we're in the first revolution [orbit]... what are you guys going to do?" Countless meetings ensued as they hammered out the meticulous details for all phases of the mission. They spent most of the time on scenarios where things went wrong.[37]

His Mission Techniques meetings began in October 1967. After each session, he summarized their conclusions by writing entertaining "Tindallgrams" which all participants looked forward to reading. Murray and Cox said, "Who else but Tindall would, in an official NASA

[35] Murray and Cox, *Apollo*, 292–293.
[36] Steve Bales, interview by author, October 9, 2019.
[37] Murray and Cox, *Apollo*, 295.

communication, describe the magnitude of a required change in spacecraft velocity as 'teensy weensy?'"[38]

A prime example of a Tindallgram came in late 1968 when he was concerned about a warning light in the LM cockpit which indicated a low fuel level:

> As you know, there is a light on the LM dashboard that comes on when there is about two minutes worth of propellant remaining in the DPS tanks with the engine operating at quarter thrust. This is to give the crew an indication of how much time they have left to perform the landing or to abort out of there. It complements the propellant gauges. The present LM weight and descent trajectory is such that this light will always come on prior to touchdown. This signal, it turns out, is connected to the master alarm—how about that! In other words, just at the most critical time in the most critical operation of a perfectly nominal lunar landing mission, the master alarm with all its lights, bells, and whistles will go off. This sounds right lousy to me. In fact, Pete Conrad tells me he labeled it completely unacceptable four or five years ago, but he was probably just an ensign at the time and apparently nobody paid any attention. If this is not fixed, I predict the first words uttered by the first astronaut to land on the Moon will be, "Gee whiz, that master alarm certainly startled me."[39]

When the Apollo 12 mission was tasked with making a pinpoint landing near the unmanned Surveyor III, he titled the ensuing memo "How to land next to a Surveyor: a short novel for do-it-yourselfers." It described "a three-day Mission Techniques free-for-all starting July 30 to see what we could jury-rig together to improve our chances of landing next to the Surveyor."[40]

The Mission Techniques meetings were similar to the CCB meetings as they both decisively accomplished issues. Yet the tones were utterly different. Flight controller Steve Bales tells of one of the

---

[38] Murray and Cox, *Apollo*, 296.
[39] Bill Tindall, "DPS low-level propellant light," Tindallgram, May 29, 1969, http://www.collectspace.com/resources/tindallgrams/tindallgrams02.pdf.
[40] Bill Tindall, "How to land next to a Surveyor: a short novel for do-it-yourselfers," Tindallgram, August 1, 1969, http://www.collectspace.com/resources/tindallgrams/tindallgrams02.pdf.

meetings he attended before Apollo 7. Bill Tindall wanted to discuss procedures for aligning the Service Module's Service Propulsion System (SPS) engine thrust through the vehicle's center of gravity. He opened the conversation by asking, "Who knows how the SPS works?" Many knew a lot about the SPS but were hesitant to speak in front of their peers. Everyone realized if their facts were not exactly right, they would get blasted from all directions.

The irrepressible Retrofire Officer John Llewellyn knew how to flush them out. He strode to the front of the room where the blackboard was located, pole position for making an argument. Llewellyn was familiar with the SPS engine and spoke confidently as he wrote on the board. Some details were accurate, and the rest were made up spontaneously as he worked the crowd. After a few minutes, an engineer in the room stood and shouted, "That's wrong! What you said is wrong," and quickly listed several objections. Llewellyn turned to Tindall and said, "Bill, there is your SPS expert." The room erupted in laughter. Steve Bales said, "Nobody would have done what Llewellyn did at a CCB. They probably would have been fired."[41]

Apollo 1 crew on the steps of the Apollo Mission Simulator at Cape Kennedy. From bottom to top, Grissom, White, and Chaffee. Photo credit: NASA 67-H-103, Kipp Teague, Ed Hengeveld.

---

[41] Steve Bales, interview by author, November 6, 2019.

*January 27, 1967. Apollo 1 astronauts approaching the White Room
and Command Module. Photo credit: NASA KSC 67PC-17, Kipp Teague.*

*Casket of Gus Grissom at Arlington National Cemetery, January 31, 1967.
Grissom is escorted by fellow astronauts (from left) Alan Shepard, John
Glenn, Gordon Cooper, and John Young. He is buried beside Roger Chaffee.
Photo credit: NASA 67-H-141, Kipp Teague, scan by Ed Hengeveld.*

7

EXCESS ELEVEN

SCIENTIST-ASTRONAUTS

On August 11, 1967, NASA announced the addition of its sixth group of astronauts. These men came from a scientific background compared to the typical test pilot route. Several months earlier, Joe Allen was finishing his PhD at Yale on his way to a career as a university physics professor when he saw a notice about NASA hiring scientist-astronauts. He was more interested in the process than becoming an astronaut, so he applied.

NASA contacted Allen in January of 1967. He progressed through the initial stages and was called to San Antonio for physical exams and then to Houston for final interviews. He was thrilled to meet people he had read about, such as Deke Slayton and Alan Shepard, which made the whole effort worthwhile. A few weeks later, he was in a physics lab when he received a message over the intercom saying he had a telephone call from a "Captain Shepard." Joe Allen made his way to the phone. "Joe, this is Al Shepard calling. We, NASA, are going to make an announcement of a class of astronauts. We want to make it at three o'clock this afternoon Houston time, and we would like your name to be on that list. Will you accept this job?" He accepted it immediately. Then he asked his wife.[1]

---

[1] Joseph P. Allen, interview by Jennifer Ross-Nazzal for NASA Johnson Space Center Oral History Project, January 28, 2003, 1–6.

Tony England was also in the new group of scientist-astronauts. Selected at 25, he remains the youngest astronaut ever chosen by NASA. His family moved to Fargo, North Dakota, when England was 12 years old. He loved carpentry and helped his father construct their home. He also built his amateur radio equipment.

England graduated high school a year early and began his studies at the Massachusetts Institute of Technology. By the age of 22, he had earned a Bachelor of Science degree in Earth sciences, plus a Master of Science degree in geology and geophysics. This is even more impressive since he did not come from a family of academics. At the time, only one other relative had earned a college degree, and that was his grandmother's sister. He was finishing his PhD in geophysics at MIT when NASA announced their recruiting of scientist-astronauts.[2]

Joe Allen, Tony England, and the other scientist-astronauts showed up for their first day of work on September 18, 1967. Deke Slayton called them in for a meeting. Slayton was feeling the pressure of congressional budget cuts. When the selection process started the previous year, NASA was flying high with the success of Gemini. The ambitious plans included up to six Apollo lunar flights per year through the mid-1970s and an Apollo Applications program to follow.[3] By the spring of 1967, congress had slashed the NASA budget. By late summer, Slayton knew many of the desired missions would be canceled.

Slayton was brutally honest as he met with the astronauts. They were not going to fly. There was useful work to be done, and if they wanted to remain in the space program they would receive interesting assignments, but they should not expect to fly. "If any of you feel that you have more important work to do elsewhere, you will make no enemies by resigning."[4]

In light of the surplus of astronauts, the 1967 class named themselves XS-11. They began five months of intense academic training, including space flight dynamics, rocket propulsion, planetology, Earth resources, space sciences, life sciences, and astronomy. The class also went through a familiarization process with

---

[2] David J. Shayler and Colin Burgess, *NASA's Scientist-Astronauts*, (Chichester, UK: Springer-Praxis, 2007), 131–134.
[3] Slayton and Cassutt, *Deke!*, 201.
[4] Joe Allen, interview by Jennifer Ross-Nazzal, 8.

the Apollo spacecraft, MOCR operation, and briefings on the Mercury, Gemini, and Apollo programs. They spent time in Huntsville getting to know the launch vehicle and at Cape Kennedy learning about launch operations.[5]

FLIGHT TRAINING

The scientists were highly motivated to prove they belonged in the astronaut corps. They began 53 weeks of Air Force flight training at different locations around the country. Because of the Vietnam war, there were many training bases in operation. They drew their assignment out of a hat.

Tony England drew Laughlin Air Force Base in Del Rio, Texas, quickly realizing it was an excellent assignment. The 23 flight school students at Laughlin developed strong bonds since there was nothing else to do in Del Rio. England considers it one of the most enjoyable years of his life.[6] He recalls, "My Air Force flight school began in a contractor-operated T-41, a Cessna 172 with a 180 HP engine, at the local airport in Del Rio, Texas. We called it the washing machine because the objective was to wash out quickly those who really did not want to fly before incurring the expense of training in jets." He flew the T-41 until he soloed, then moved to the Cessna T-37, a small twin-engine jet trainer. After logging 120 hours in the T-37, he stepped up to the T-38 for his final 100 hours. England continues,

> We graduated with 250 hours in 53 weeks. We then flew NASA's modified T-38s under restrictions until we had built 400 hours. After that, we were free to fly anywhere and as often as any of the other astronauts flew. Some of the pilot-astronauts routinely broke rules that we scientist-astronauts knew we could not break. Consequently, they occasionally had accidents we knew we could not have had and remain in the program.[7]

---

[5] Shayler and Burgess, *NASA's Scientist-Astronauts*, 171–173.
[6] Tony England, "Human Spaceflight: A History of Competing Objectives," Mardigian Library Lecture Series, (lecture, University of Michigan-Dearborn, February 15, 2019), accessed July 1, 2019, video, 1:36:50, https://library.umd.umich.edu/lecture/.
[7] Tony England, email message to author, December 22, 2021.

Eight of the eleven scientist-astronauts made it through flight training. England and Allen were typical of those eight, graduating at the top of their flight training class. Allen loved his experience as well. "I had a new kind of classroom. I went in being unable to fly and came out fifty-three weeks later as a very-highly-qualified pilot." He rated first in all five graded categories: academics, general flying, instrument flying, acrobatics, and formation flying. He acknowledges being older with more wisdom gave him an advantage over the other students. "Flying, to a degree, is a mind discipline game more than it is a physical athlete's game."[8]

It is no surprise scientist-astronauts excel in flight training. They have years of experience taking new information, thoroughly engaging it, and applying it to professional tasks. Most flight students come from a college background where they memorize information to pass a test and then move on. US Navy flight instructor Jeff Hale explains, "The astronauts do not simply put knowledge into their brain, but they are able to process it, explain it, teach it, and understand the concepts behind it in such a way that is so far superior to other flight students that it gives them a significant advantage." Their experience benefits them in briefings and inside the cockpit. For many 23-year-old students, flight training is the first time they've ever had to internalize information and perform procedures within various constraints. But that does not phase the astronauts.[9]

Tony England adds, "We tend to be more mature. We finished our PhD degrees. We're older. When you are an undergraduate, you take courses for grades. But someplace along the line you start taking courses to learn. And that's a completely different way to learn." England went to MIT on a merit scholarship and had to maintain high grades. "I went in fear of not getting good-enough grades for three years. My senior year I knew they weren't going to throw me out, so I relaxed, and my grades went up because I started learning as I should have, instead of learning for grades."[10]

Good judgment honed by positive professional experience is another area that separates scientist-astronauts from younger flight

---

[8] Joe Allen, interview by Jennifer Ross-Nazzal, 11.
[9] Jeff Hale, interview by author, July 12, 2019.
[10] Tony England, interview by author, May 19, 2021.

students. The astronauts have successfully navigated obstacles to reach the top of their field and would not have been selected by NASA if their judgment was less than stellar. Therefore, they avoid putting themselves in dangerous situations in the aircraft. Jeff Hale referred to a quote by Frank Borman which appears on many Naval Aviation posters: "A superior pilot uses his superior judgment to avoid situations which require the use of his superior skill."[11]

The veteran astronauts noticed the excellent work of their new comrades. Joe Allen received a letter that he still treasures from Alan Shepard, who was called "Ice Commander" by many in the astronaut office. "Dear Joe, I realize what you've done. You have made all of us at NASA very proud." Allen's wife wasn't so easily impressed. Returning home, he boasted, "Bonnie, I came home with a trophy!" She said, "Well, it's about time. You don't even have a bowling trophy."[12]

MORE TRAGEDIES

1967 saw more tragedies in the space program. Astronaut Ed Givens died in a car accident near Houston on June 6. He was the first of the class of 1966 to receive a crew assignment for an Apollo flight. C.C. Williams died on October 5 when his T-38 suffered a mechanical failure and crashed near Tallahassee, Florida. He was slated to be the Lunar Module Pilot on Pete Conrad's crew. Al Bean replaced Williams and walked on the Moon with Conrad on Apollo 12.

On November 15, X-15 pilot Mike Adams perished when his plane spun out of control and broke apart after reaching 266,000 feet. The US Air Force posthumously awarded Adams his astronaut wings since he reached their defined boundary of space of 50 miles (264,000 feet).

Robert Lawrence, an accomplished flight instructor at Edwards and a member of the Air Force Manned Orbiting Laboratory program, was killed when a student crashed an F-104 jet in December. The MOL was canceled two years later, and seven of its astronauts transferred to NASA.[13] Lawrence was the first African American to qualify as an

---

[11] Jeff Hale, interview by author, July 12, 2019.
[12] Joe Allen, interview by Jennifer Ross-Nazzal, 12.
[13] All seven eventually flew on Space Shuttle missions.

astronaut. Deke Slayton said he definitely would have been accepted by NASA with the seven others.[14]

The Russian space program suffered its own tragedy on April 24, 1967. Vladimir Komarov was flying the first manned Soyuz mission when his spacecraft suffered a parachute failure, impacting the ground at high speed. Komarov was killed instantly. His ashes were interred in the Kremlin wall during his state funeral in Moscow.

TURNING POINT

The most visible sign of Apollo's development came on November 9, 1967, when the world witnessed the first launch of a Saturn V. It was daring, dramatic, and successful.

Wernher von Braun and his team were conservative in their testing, introducing only one new element into a flight. When George Mueller became Director of the Office of Manned Space Flight in September of 1963, he immediately realized all the methodical unmanned launches would run NASA over budget and well past the end-of-decade deadline for a lunar landing. He relied on his extensive experience leading Air Force missile programs where an all-up concept was introduced with the Titan II missile, testing all the components in one flight.[15]

He applied this method to the Saturn V; all three stages of the launch vehicle would be live and tested simultaneously in a risky make-or-break move. Not only was it a trial for the launch vehicle but also an evaluation of the Block II heat shield attached to an unmanned Block I Command Module.

At least ten unmanned flights of the Redstone, Atlas and Titan launch vehicles were flown before they were approved for manned Mercury or Gemini missions. However, years of meticulous ground testing gave confidence that they only needed two unmanned Saturn V launches. The ambitious first test flight was called Apollo 4.[16] In *Chariots for Apollo*, Brooks, Grimwood, and Swenson paint the picture:

---

[14] Slayton and Cassutt, *Deke!*, 206.

[15] *See* pages 18–19 for further explanation. *See also* Bilstein, *Stages to Saturn*, 348–349.

[16] The wives of Gus Grissom, Ed White, and Roger Chaffey requested the Apollo 1 designation be used for their husbands' mission. There were also unmanned tests launched on the smaller Saturn 1B before the fire.

Birds, reptiles, and animals of higher and lower order that gathered at the Florida Wildlife Game Refuge [also known by the aliases of Merritt Island Launch Annex and Kennedy Space Center] at 7:00 in the morning of November 9, 1967 received a tremendous jolt. When the five engines in the first stage of the Saturn V ignited, there was a man-made earthquake and shockwave. As someone later remarked, the question was not whether the Saturn V had risen, but whether Florida had sunk.[17]

It is said the only man-made sounds louder than a Saturn V launch are nuclear explosions.[18] Over three miles away, CBS news correspondent Walter Cronkite had the enthusiasm of a small child watching the historic liftoff. As the mighty rocket cleared the launch tower, the sound waves pounded his broadcast booth. "Our building is shaking here!! Our building is shaking!! The ROAR is terrific!! The building is shaking! This big glass window is shaking and we're holding it with our hands. Look at that ROCKET go!!" He exclaimed these memorable words while being showered by debris from the quaking walls and ceiling.

The five first-stage engines produced 7.5 million pounds of thrust. During the next two and a half minutes, they propelled the rocket to a speed of over 8,800 feet per second at an altitude of 35 nautical miles, 45 nm downrange.[19] Engineering cameras on the second stage captured the dramatic scenes of the first stage separating and falling away. The cameras also filmed the release of the inter-stage skirt, which connected the first and second stages. The video was important to the engineers because there was minimal clearance around the second-stage engines. The cameras later parachuted into the Atlantic Ocean, were successfully recovered, and the highly-anticipated film gave engineers important feedback. Many documentaries have used the unforgettable images.

The S-II second stage, built by North American Aviation, had been troublesome for years. The welding processes were complicated, plus

---

[17] Brooks, Grimwood, and Swenson, *Chariots for Apollo*, 232.
[18] Bilstein, *Stages to Saturn*, 357.
[19] NASA, "Saturn V Launch Vehicle Flight Evaluation Report: AS-501 Apollo 4 Mission," (Huntsville: George C. Marshall Space Flight Center, 1968), https://archive.org/details/nasa_techdoc_19900066482/page/n85/mode/2up, 4-5.

there was a growing need to shave weight from the thin-walled S-II. North American restructured its management, and NASA sent technical teams to assist in the S-II development. The second stage arrived at the Cape in January of 1967, but examinations discovered hairline cracks causing the entire stage to undergo a thorough X-ray inspection. Other areas of concern needed intricate repairs and added to the delays.[20] It was time to see how it performed under the enormous stresses of powered flight.

The S-II engines started flawlessly and burned for over six minutes, carrying the rocket to a speed of 22,000 feet per second at an altitude of 104 nautical miles, 800 nm downrange. It passed the torch to the third stage S-IVB, which burned for over two minutes to reach the orbital velocity of 25,500 fps.[21]

Not only did the launch vehicle pass the test, but so did the Block II heat shield. Flight Director Glynn Lunney said, "The idea there was to propel the vehicle all the way to conditions just about what they would reach when they reentered from the Moon. We came close to those kinds of conditions."[22]

This was performed by igniting the third stage again to send the spacecraft into an elliptical orbit with a maximum altitude of 9,300 nautical miles. From there, the mated Command and Service Modules (Command/Service Module, or CSM) separated from the booster and was aimed toward Earth. The Service Module engine fired for almost five minutes, driving the CSM toward the atmosphere at 36,500 feet per second, simulating a return from the Moon.[23]

The Command Module separated from the Service Module and turned around with its blunt heat shield leading it into a fiery entry. The CM survived and landed safely in the ocean within 10 miles of its recovery ship. The mission accomplished every technical goal and was a complete success, launching the Apollo program into 1968 with great optimism and momentum.

---

[20] Bilstein, *Stages to Saturn*, 231–232.
[21] NASA, "Saturn V Launch Vehicle Flight Evaluation Report: AS-501 Apollo 4," 4-5.
[22] Glynn S. Lunney, interview by Carol Butler for Johnson Space Center Oral History Project, February 8, 1999, 3.
[23] Brooks, Grimwood, and Swenson, *Chariots for Apollo*, 232–233.

NASA Astronaut Group Six, 1967, nicknamed the XS-11.
Joe Allen is on the back row, far left, beside the flag.
Tony England is on back row, third from left. Photo credit: NASA.

Apollo 4. The first Saturn V all-up test flight. Engineers installed
onboard cameras to see close-up views of staging events since the clearances
between engines and structures were extremely tight.
Screen capture from NASA film.

SATURN V SPECIFICATIONS

INSTRUMENT UNIT (IU)

Diameter:	21.7 ft.
Height:	3 ft.
Weight:	4,500 lbs.

THIRD STAGE (S-IVB)

Diameter:	21.7 ft.
Height:	59.3 ft.
Weight:	260,000 lbs. fueled
	25,000 lbs. dry
Engine:	One J-2
Propellants:	Liquid Oxygen
	(189,800 lbs.; 20,000 gals.)
	Liquid Hydrogen
	(43,500 lbs.; 74,150 gals.)
Thrust:	198,800 lbs. to 230,000 lbs.
Interstage:	8,000 lbs.

SECOND STAGE (S-II)

Diameter:	33 ft.
Height:	81.5 ft.
Weight:	1,101,000 lbs. fueled
	78,000 lbs. dry
Engines:	Five J-2
Propellants:	Liquid Oxygen
	(837,200 lbs.; 88,200 gals.)
	Liquid Hydrogen
	(159,700 lbs.; 272,200 gals.)
Thrust:	1,150,000 lbs.
Interstage:	11,400 lbs.

FIRST STAGE (S-IC)

Diameter:	33 ft.
Height:	138 feet lbs.
Weight:	4,930,000 fueled
	289,800 lbs. dry
Engines:	Five F-1
Prepellants:	Liquid Oxygen
	(3,306,000 lbs.; 348,300 gals.)
	RP-1 Kerosene
	(1,438,000 lbs.; 215,700 gals.)
Thrust:	7,766,000 lbs. at lift-off

APOLLO
SPACECRAFT

IU

S-IVB
STAGE

363 FEET

S-II
STAGE

S-IC
STAGE

NASA-JSC

The mighty Saturn V launch vehicle with the Apollo spacecraft on top.
The spacecraft consisted of the Command Module, Service Module, Lunar
Module, plus the Spacecraft-Lunar Module Adapter and Escape Tower.
Artwork credit: NASA.

8

THE YEAR THAT SHATTERED AMERICA

1968

The calendar turned to 1968, twelve months full of political turmoil and social unrest. The Smithsonian Magazine calls it "The Year That Shattered America."[1] In March, President Lyndon Johnson announced his name would not be on the ballot for the November Presidential election. Half a million troops fought in Vietnam. Racial tensions were at a fever pitch. Dr. Martin Luther King, Jr. was assassinated in April. Riots erupted in over 100 cities throughout the country. Senator Robert Kennedy, a popular candidate in the Presidential race, was assassinated in June. The Democratic National Convention was held in Chicago while 10,000 anti-war protesters clashed with over 20,000 Police and National Guardsmen outside the convention center.[2] America needed something good to happen. NASA delivered—but it was not without anxious moments.

In January 1968, engineers evaluated the first Lunar Module during a 21-day unmanned flight. The launch vehicle was the Saturn 1B originally designated for Apollo 1. Since the LM had no heat shield, it suffered a fiery entry and its debris fell into the Pacific Ocean. It was a

---

[1] Matthew Twombly, "A Timeline Of 1968: The Year That Shattered America," *Smithsonian Magazine*, January/February 2018,
https://www.smithsonianmag.com/history/timeline-seismic-180967503/.
[2] For a summary of the significance of 1968 in the story of Apollo 8, *see Rocket Men* by Robert Kurson, chapter 13.

successful test, although weight would have to be trimmed before a manned version could fly to the Moon. Also, the LM ascent engine was still plagued by nagging fuel injector issues.

On Thursday, April 4, the second and final unmanned Saturn V test thundered off the launch pad at 7:00 a.m. The mission was designated Apollo 6, and in the words of Glynn Lunney, "that flight really misbehaved."[3]

Two minutes after liftoff, severe vibrations up and down the launch vehicle emerged, resembling a pogo stick. The natural vibrations of each engine's thrust chamber and combustion chamber occurred about 5.5 times each second, or a frequency of 5.5 hertz. A "pogo task force" later concluded the engine frequency too closely matched the vibration rate of the whole vehicle (5.25 hertz) late in the burn. For future missions, the engines were detuned to avoid the same frequencies as the launch vehicle.[4]

Glynn Lunney described this problem in layman's terms. "It's a little bit like the soldiers marching across a bridge in cadence. We learned not to do that, too, because it's hard on bridges and people who fall down when they do! And stages could get to vibrating up and down, mostly longitudinally, and really destroy themselves."[5]

A second way to avoid the pogo issue was to shut down the main culprit, the center engine, before it reached the point where the weight and thrust combination produced the pogo-inducing frequency. "Then the other four engines just continued to burn the propellant, and we got the same amount of energy out of the stage than we otherwise would have gotten. We just got to it a little more slowly than we otherwise would," said Lunney.[6]

The first stage was not the only problem on Apollo 6. After the first stage depleted its propellants, the second stage took over. Two of the five J-2 engines quit without warning a few minutes early. Typically, it would have caused the booster flight controller Bob Wolf to hit the abort button, but he decided to carry on since there were no signs of instability.

---

[3] Glynn Lunney, interview by Carol Butler, February 8, 1999, 3.
[4] Bilstein, *Stages to Saturn*, 362–363.
[5] Glynn S. Lunney, interview by Roy Neal for NASA Johnson Space Center Oral History Project, April 26, 1999, 18.
[6] Glynn Lunney, interview by Roy Neal, April 26, 1999, 18.

The other three engines had to fire longer than planned. Afterward, the third-stage engine took the rocket into orbit but failed to reignite when it was time to simulate the burn sending the spacecraft to the Moon.

The problems encountered during the test flight were overshadowed by events in Memphis later in the day. Dr. Martin Luther King, Jr. was murdered on the second-floor balcony of the Lorraine Motel. The terrible year had become much worse.

PROGRESS

The following morning, Jim Lovell, Charlie Duke, and Stu Roosa climbed aboard Command Module 007 for a two-day voyage in the Gulf of Mexico to test its seaworthiness. Could the Block II CM float for an extended time if recovery ships could not arrive immediately after splashdown?

Their test included an evaluation of the Command Module's ability to right itself if tipped over in a nose-down attitude, called the "Stable II" position. Closely watched by a recovery ship, the CM experiment began with its nose underwater and blunt end toward the sky. The crew inflated three balloon-like devices in the nose to upright the spacecraft. Once in the standard "Stable I" position, it was a matter of waiting for the end of the test.

Charlie Duke quickly remembered why he chose flying over being at sea. It was a hot, sticky, miserable 48 hours bobbing in the water, forcing him to take motion sickness pills. "The biggest mistake I ever made was volunteering for that deal," said Duke. "Lovell was so nice and said he needed two guys to help him. So Stu and I agreed and thought we would get a little experience inside the spacecraft."[7]

To their surprise, there were no flight systems on board; it was only a boilerplate. The only food supplies were typical rations remaining after a lunar voyage, plus some fecal bags and urine bags. "It was basically a test of the Environmental Control System," Duke explained. "You were sucking in humid sea air which condensed all over the walls. The sweat from the walls got all over you, and it was salty and miserable. We were

---

[7] Charlie Duke, interview by author, May 20, 2020.

really glad to get out of there. All three of us wondered why we volunteered."[8]

Afterward, the crew made their official report. Although they "did not recommend the Apollo spacecraft for any extended sea voyages, they encountered no serious habitability problems during the 48-hour test. If a comparison can be made, the interior configurations and seaworthiness make the Apollo spacecraft a much better vessel than the Gemini spacecraft."[9]

Duke was appointed to monitor Houston's work with the Marshall Space Flight Center in Huntsville as they addressed the pogo problems the Apollo 6 flight encountered. A Saturn V manager estimated a thousand engineers were working on the problem at one point. Charlie Duke spent a lot of time in Huntsville relaying updates to Houston.[10]

Duke was still heavily involved with the special committee overseeing the LM ascent engine injector. Any further delays threatened to push the entire Apollo program behind schedule. Even minor changes like new welding techniques caused chaos with quality control when trying to determine why some engine components failed qualification tests. Apollo Program Director Sam Phillips told George Low, "The ascent engine was one of a very few Apollo hardware items in which even the most insignificant change must be elevated to top-level management review before implementation."[11]

On May 28, NASA met with Grumman officials to decide which injector to use in the LM ascent engine. They discussed whether to stay with the Bell injector or change to the Rocketdyne version late in the game. Low went around the room, asking each person for their recommendation. "I was the last guy," recalled Charlie Duke. Low asked him, "What do you want to do?" Duke replied, "We need to go with Rocketdyne." Low said, "We're going with Rocketdyne."[12] In *The Apollo Spacecraft: A Chronology,* Ertel, Newkirk, and Brooks summarize, "After long months of technical effort and almost agonizing

[8] Charlie Duke, interview by author, May 20, 2020.
[9] Ivan D. Ertel and Roland W. Newkirk, *The Apollo Spacecraft: A Chronology, Volume IV, January 21, 1966-July 13, 1974*, with Courtney G. Brooks, (Washington, DC: National Aeronautics and Space Administration, 1978), 217.
[10] Brooks, Grimwood, Swenson, *Chariots for Apollo*, 251–252.
[11] Ertel, Newkirk, and Brooks, *The Apollo Spacecraft Chronology, Volume IV*, 224.
[12] Charlie Duke, interview by author, May 20, 2020.

hardware and managerial debate, the issue of an ascent engine for the LM was settled."[13]

By the end of May, Frank Borman reported that the redefined Command/Service Module was almost ready to fly to the Moon. The first CSM would be shipped to the Cape ahead of schedule. "The CSM was in the best shape of any spacecraft ever when it arrived in Florida," and went through the various checks "with flying colors," said Chris Kraft.[14] Low, who enjoyed James Bond films, sent what he referred to as a "007 memo" to Kraft, who was to destroy it after reading the message. It asked him to study the issues involved with a lunar mission from a flight operations standpoint.[15]

The Russians were getting close to a lunar mission, only months away from sending an unmanned Zond spacecraft around the Moon. In late July, Low thoroughly reviewed the Lunar Module's progress and was deeply discouraged by his conclusion. Any hopes for the LM to be flight-ready by the end of the year were gone.

The delay did not affect the first manned Apollo flight, Apollo 7, a test of the Command/Service Module in October. However, the Lunar Module issue would directly alter the following mission commanded by Jim McDivitt, which would evaluate the CSM and LM in low Earth orbit. Next would be Apollo 9, commanded by Frank Borman, repeating the assessment in a much higher orbit. Low went on a Caribbean vacation but could not get the problems off his mind. "That is when George Low showed his genius," said Chris Kraft.[16]

Low feared waiting for the LM would give the Russians a golden opportunity to win the race to the Moon. What if they shuffled Apollo flights allowing a December flight around the Moon with the Command/Service Module while McDivitt's mission waited for the Lunar Module? A lunar flight would require the mighty Saturn V booster which only had two unmanned test flights under its belt. The idea of manning the next Saturn V was not unheard of; in April, Apollo

[13] Ertel, Newkirk, and Brooks, *The Apollo Spacecraft Chronology, Volume IV*, 228.
[14] Kraft, *Flight*, 284.
[15] Murray and Cox, *Apollo*, 315.
[16] Kraft, *Flight*, 284.

Program Director General Sam Phillips recommended to NASA Administrator James Webb a manned Saturn V be flown in late 1968.[17]

When Low returned to Houston, he talked about adjusting the schedule to his assistant Scott Simpkinson who was in charge of flight safety. Simpkinson had no concerns from a spacecraft point of view; it was mainly a navigation issue. Its subject matter expert was Bill Tindall, who responded to the idea like gangbusters. His Mission Techniques meetings had already worked out many navigation procedures and would be able to finish the remaining details in time.[18]

Once Low knew it would be technically possible, he had to ask another question—was it crazy? Apollo missions required at least 18 months of intense planning and up to three years to prepare for the more complex ones.[19] They had four months to pull this off and only a few days to decide. NASA could not afford any additional mistakes. The Apollo program almost died after the fire 18 months earlier; one more accident could doom it. The costs would be high, but the potential gains would be even higher.

Low met with Kraft and Bob Gilruth and got straight to the point. "What do you think about flying to the Moon without a Lunar Module? If Apollo 7 does okay in October, can we juggle the schedule to fly a crew to the Moon and back in December?" Kraft and Gilruth were shocked but wanted to hear more. NASA did things incrementally, one step at a time. This plan to bypass several of those steps was audacious to say the least. Low continued to make his case, "It would ace the Russians and take a lot of pressure off Apollo. And we have to go there sooner or later anyway."[20]

They called in Deke Slayton, who often joined Low, Kraft, and Gilruth for discussing off-the-wall ideas. Low asked if a crew could be ready in four months. It would take crew swapping, so Slayton told them he needed a day or two to analyze it and make some calls.

After the meeting, Kraft assembled a group of his best mission planners, including his technical assistant Rod Rose, John Mayer, head

---

[17] NASA, *Roundup*, Vol 7, No 14, April 26, 1968, (Houston: Manned Spacecraft Center Public Affairs Office, 1968), 1.
[18] Murray and Cox, *Apollo*, 317.
[19] NASA, "What Made Apollo a Success, Updated August 5, 2004," https://history.nasa.gov/SP-287/sp287.htm, 70.
[20] Kraft, *Flight*, 284.

of the Mission Planning and Analysis Division, and Bill Tindall. Low's idea called for the crew to be sent around the Moon on a "free return trajectory," using lunar gravity to slingshot the spacecraft back to Earth with only a few midcourse corrections. The concept of free return trajectory started several years earlier and was well known to them.

"Their eyes lit up like Christmas trees, particularly Mayer's and Tindall's," said Kraft. "This was a challenge they wanted. They were gung-ho guys, willing to do whatever it took. I sent them off to look at everything from getting Mission Control ready to having operational trajectory software which would take three Americans around the back of the Moon and bring them home again."[21]

They reported back to Kraft the next morning. The mission could be done, and they had an even bolder suggestion. Since they were going to the Moon, why not go into lunar orbit and simulate the same orbit the landing missions would use? It would help them better understand some of the unusual orbital characteristics they noticed when unmanned spacecraft circled the Moon.

Kraft loved what he heard and called Low, who arranged a meeting with Slayton and Gilruth the next day. Kraft's team prepared a detailed package showing how it could work. Everyone supported the plan, and it was time to get feedback from two more men.

They walked out of Gilruth's office and boarded a plane to Huntsville, Alabama. A few hours later, they bounced their idea off of Wernher von Braun and General Sam Phillips, who directed the Apollo Manned Lunar Landing Program. Von Braun and General Phillips were excited. Phillips told the group, "This is the best idea I've heard yet. I'm proud of you guys for coming forward with this."[22]

On Sunday, August 12, Ken Mattingly was in Downey, California, assisting with tests on the Apollo 9 Command Module.[23] The mission would be a high-Earth orbit repeat of the preceding flight by Jim McDivitt's crew. The Apollo 9 prime crew was also present: Commander Frank Borman, Jim Lovell, and Bill Anders.

---

[21] Kraft, *Flight*, 285.
[22] Kraft, *Flight*, 287.
[23] Mattingly was assigned to their Support Crew after finishing his work supervising and testing the Apollo spacesuit and Portable Life Support System (PLSS).

In the middle of a test, Borman received an urgent phone call from Deke Slayton telling him to drop everything and return immediately to Houston. Borman had no idea what could be that important, but he followed orders while Mattingly and the other astronauts stood by in California. Borman landed his T-38 at Ellington Field, then drove straight to Slayton's office. Slayton told him about the daring plan and asked if his crew would be interested in flying the new mission. Borman, extremely motivated to beat the Russians, immediately said yes. He flew back to California to update his crewmates.

Jim Lovell was elated. "I thought, 'Man, this is great!' I had already spent two weeks in space in Gemini VII with Frank Borman. I didn't want to spend another 11 days going around Earth again."[24]

Bill Anders had mixed feelings. As a patriot, he knew the importance of beating Russia to the Moon. But he was a Lunar Module Pilot, in line for a lunar landing on a later flight. Before the 1967 fire, he trained in the Lunar Landing Training Vehicle (LLTV) along with Neil Armstrong, which he took as a sign of good things to come. The new plan for Apollo 8 removed the LM from their flight, putting Anders on track to become a Command Module expert. Any hope of landing on the Moon would now depend on him being a Commander, and his low ranking on the NASA totem pole meant it would not happen for many years. Anders later reflected, "I feel extremely fortunate to have been able to participate on man's first flight away from our own planet, but I would have traded the last lunar landing for the first flight away from the planet."[25]

Deke Slayton's crew swapping involved Jim McDivitt and his men scheduled to fly after Apollo 7. McDivitt's mission would be what every test pilot longed for, flying the Lunar Module for the first time and putting the Command/Service Module through every scenario imaginable. McDivitt, Dave Scott, and Rusty Schweickart had already trained a year and a half for their mission, and Slayton did not want their experience wasted. Their mission would now be called Apollo 9. McDivitt recalls, "Over the years this story has grown to the point where

---

[24] James A. Lovell, Jr., interview by Ron Stone for NASA Johnson Space Center Oral History Project, May 25, 1999, 27.

[25] William A. Anders, interview by Paul Rollins for NASA Johnson Space Center Oral History Project, October 8, 1997, 9–10.

people think I was offered the flight around the Moon but turned it down. Not quite. I believe if I'd thrown myself on the floor and begged to fly the new mission, Deke would have let us have it. But it was never really offered."[26]

A flurry of activity ensued. Mission planners, flight controllers, and the crew had to funnel years' worth of work into four short months preparing for the new Apollo 8 mission. Ken Mattingly compared it to "cramming for a degree.... It was 24 hours a day. You lived it. You grudgingly would grab a couple hours of sleep."[27]

Planning the mission on short notice was a challenge. But the actual flight would be a heart-stopper, incorporating many elements new to manned spaceflight. The launch would be the first time a human rode atop the mighty Saturn V launch vehicle. Another challenge was entering Earth's atmosphere from a lunar distance, traveling faster than man had ever flown. And in between was possibly the most daunting part of the mission, entering and exiting lunar orbit.

Those maneuvers required successful burns of the Service Module's Service Propulsion System (SPS) engine. The burn for entering lunar orbit would be a braking maneuver to slow the CSM enough to be captured by lunar gravity. During the final lunar orbit, the engine would have to reignite successfully to send the crew away from the Moon and back to Earth. If the latter of those burns did not happen perfectly, the men would be stranded. Bill Anders calculated there would be a 33 percent chance for a successful mission, a 33 percent chance they would abort and come back alive without making it to the Moon, and a 33 percent chance they would not survive.

RUSSIANS UP THE ANTE

On September 18, 1968, British scientist Sir Bernard Lovell reported the Soviet unmanned spacecraft Zond 5 passed within 1,000 miles of the Moon and was heading back to Earth. The Russians confirmed it the following day. After the spacecraft returned successfully, Sir Bernard Lovell stated, "It makes it highly probable that a Russian will get a close-up look at the Moon quite a long time before an American does." The

---

[26] Slayton and Cassutt, *Deke!*, 215.
[27] Ken Mattingly, interview by Rebecca Wright, November 6, 2001, 24.

expert on the Soviet space program continued, "I think it is a very considerable achievement and I expect that a human being will be placed in a similar spacecraft in a matter of months."[28]

The unmanned Zond 6 flew around the Moon on November 14. The *New York Times* reported:

> Informed sources said this week that the Russians were preparing to send at least two and probably three men on a flight around the Moon "soon." These sources said the flight could come any time within the next two weeks and might be undertaken before the launching of the American Apollo 8, scheduled for December 21. Apollo 8 is to take three men on a circumlunar journey.[29]

The United States could not afford to be nipped at the finish line by the Russians again. Adding to the angst was the calendar. Celestial mechanics dictated the best time for an American lunar flight in December would be the week of Christmas. On December 24, the lighting conditions would be the same as they would for the first landing attempt on the Sea of Tranquility. The Commander flying the LM needed to have the Sun low in the sky behind him as he picked a landing spot, for long shadows highlighted any craters and boulders to avoid. The thought of losing another crew was terrible enough, but for it to happen around the Moon during Christmas would worsen a horrible year and put the space program's future in tremendous jeopardy.

John F. Kennedy realized such risks would be necessary when he challenged the nation to accomplish a lunar landing. On November 21, 1963, twenty-four hours before his assassination, President Kennedy spoke at the Brooks Air Force Base School of Aerospace Medicine dedication in San Antonio, Texas. He began his address by saying,

> For more than three years I have spoken about the New Frontier. This is not a partisan term, and it is not the exclusive property of Republicans or Democrats. It refers, instead, to this nation's place in history, to the fact that we do stand on

---

[28] "Lovell Sees Soviet Leading," *New York Times*, September 23, 1968, https://timesmachine.nytimes.com.
[29] "TASS Says Zond Craft Open Way to Manned Flight Around Moon," *New York Times*, November 30, 1968, https://timesmachine.nytimes.com.

the edge of a great new era, filled with both crisis and opportunity, an era to be characterized by achievement and by challenge. It is an era which calls for action and for the best efforts of all those who would test the unknown and the uncertain in every phase of human endeavor. It is a time for pathfinders and pioneers."

He ended the speech by telling of the Irish writer Frank O'Connor. When he was a boy, O'Connor and his buddies encountered a wall that seemed too high to climb:

They took off their hats and tossed them over the wall—and then they had no choice but to follow them. This nation has tossed its cap over the wall of space, and we have no choice but to follow it. Whatever the difficulties, they will be overcome. Whatever the hazards, they must be guarded against. With the vital help of this Aerospace Medical Center, with the help of all those who labor in the space endeavor, with the help and support of all Americans, we will climb this wall with safety and with speed—and we shall then explore the wonders on the other side. [30]

April 5-7, 1968. Command Module 007 with Lovell, Duke, and Roosa aboard. Nearby is a Motor Vessel Retriever acquired by NASA from the US Army. Photo credit: NASA, Robert Hurst.

---

[30] John F. Kennedy, "Remarks at The Dedication of The Aerospace Medical Health Center," San Antonio, Texas, November 21, 1963, JFK Library Foundation, https://www.jfklibrary.org/archives/other-resources/john-f-kennedy-speeches/san-antonio-tx-19631121.

Apollo 8 crew, from left to right: Jim Lovell, Bill Anders,
and Frank Borman. Photo credit: NASA S68-50265.

9

PAUL HANEY,
ARE YOU A TURTLE?

APOLLO 7

Apollo 7 had to be a flawless test of the Command/Service Module for a lunar mission to be flown in December. Commander Wally Schirra, a veteran of Mercury and Gemini flights, was joined by Donn Eisele and Walt Cunningham. The backup crew for Apollo 7 consisted of Tom Stafford, John Young, and Gene Cernan, who would become the prime crew for Apollo 10. The mission objective was to thoroughly evaluate the new Block II Command Module. Since no Lunar Module was ready to fly, the launch vehicle would be the Saturn 1B, an uprated version of the veteran Saturn 1 booster.

NASA added three new Flight Directors to the stable due to the enormous workloads preparing for the upcoming Apollo flights. Gerry Griffin was selected after working throughout Project Gemini as a Guidance, Navigation, and Control flight controller. Milt Windler was promoted from the spacecraft recovery division. Pete Frank had been the head of the Mission Analysis Branch, where he was responsible for the Reentry and Lunar Trajectory sections. All three had been military pilots accustomed to the pressure Flight Directors would encounter. Gerry Griffin was the first of the three to see action, assigned to the inaugural manned Apollo mission.

Apollo 7 launched from Pad 34 on a warm October 11, 1968.[1] Despite head colds suffered by the crew, it achieved all the technical goals for the mission. The astronauts gave the Apollo Command/Service Module a complete examination, proving it would meet all the demands of traveling to the Moon and back.

After achieving an orbit of 123 by 153 nautical miles, the crew used the Reaction Control System (RCS) thrusters to separate the Command/Service Module (CSM) from the booster. After turning around, they returned to the booster, simulating the future maneuver crews would use to extract the Lunar Module from the third stage. The next day they tested the Service Module engine to rendezvous with the Saturn booster 75 miles away. All went according to plan. The crew successfully fired the Service Module engine eight times during the eleven-day mission. They also tested the equipment and systems vital to the success of future Apollo missions. The few minor problems encountered were either solved during the flight or fixed for future flights.[2]

The head colds suffered by the astronauts became an issue. It is bad enough flying in an airplane with a cold, but being in space is much more painful since there is no gravity to drain the mucus from nasal passages. The flight crew did not hide their discomfort, but showed no effects of illness on the mission's fourth day as time approached for the first live television broadcast from space.

Onboard the spacecraft was an RCA black-and-white slow-scan television camera, which the crew would activate as they approached Texas. The tracking stations in Corpus Christi, Texas, and Merritt Island, Florida (near Cape Kennedy) were the only ones equipped with an RCA slow-scan converter to adjust the signals to US commercial television standards. The telecast would last seven minutes until they were past the Florida station.

Gerry Griffin's Gold Team manned the consoles for the first television broadcast. Veteran astronaut Tom Stafford, who flew with Schirra on Gemini VIA, swooped in and replaced Jack Swigert as

---

[1] NASA used Pad 34 to launch the smaller Saturn 1 and Saturn 1B rockets. The mighty Saturn V launches occurred at the Pad 39 complex.
[2] NASA, "Apollo 7 Mission Report," (Houston: Manned Spacecraft Center, 1968), https://www.hq.nasa.gov/alsj/a410/A07_MissionReport.pdf, 1-1.

CAPCOM. Dick Nafzger from Goddard Space Flight Center supported the mission in Corpus Christi. He took his portable tape recorder for the occasion, and the local team brought a television into the Operations Building. As the big moment approached, Nafzger turned on the tape recorder, capturing the comments of ABC television commentator Jules Bergman in the background along with the voices of the local team.

The technically-minded Schirra, Eisele, and Cunningham put on an entertaining show once they turned on the camera at 9:45 a.m. Houston time. Michael Kapp, a music executive who provided cassette tapes full of songs that the astronauts took into space, furnished cue cards for the broadcast. Schirra reached for the first card, which read, "From that Lovely Apollo Room High Atop Everything."

After reading the message, CAPCOM Tom Stafford commented, "The definition is pretty good down here; I can see the center hatch. Actually, I am amazed; it looks real good."

Referring to himself as "Cecil B. DeStafford," he repositioned Eisele.[3] Schirra retrieved the second sign, "Keep Those Cards and Letters Coming In, Folks."

Walt Cunningham pointed the camera out the window as they went over the Gulf Coast:

> CAPCOM (STAFFORD): Let's take a look and see how New
> Orleans is this morning.
> CUNNINGHAM: Roger. Coming up over the Mississippi River.
> I'm giving you an out-the-window picture. You should see
> Lake Pontchartrain coming into view now.
> SCHIRRA: Going over Mobile now, and we'll be coming
> across Pensacola shortly.

The broadcast was a great success. The "Walt, Wally, and Donn Show" won an Emmy Award for their seven live broadcasts during the mission. Twenty-four hours later it was time for the second highly-anticipated transmission. The screens at the front of Mission Control displayed television pictures. Jack Swigert was CAPCOM, and Dick Nafzger had his trusty tape recorder running again. The Texas tracking

---

[3] Cecil B. DeMille was a very successful film director and producer.

station gained acquisition of the spacecraft at 9:30 a.m. and could see the television images a minute later.

Eisele appeared by himself as the telecast started and held up the same opening card from the previous day. He introduced the festivities, "Coming to you live from outer space, the one and only original Apollo orbiting road show, starring those great acrobats of outer space, Wally Schirra and Walt Cunningham."

Right on cue, Schirra floated into view holding a new card. The team at Corpus Christi erupted in laughter when it appeared on the screen.[4]

> CAPCOM (SWIGERT): Looks like it says, "Are you a Turtle, Deke Slayton?"
> EISELE: You get an A for reading today, Jack.

The "Ancient and Honorable Order of Turtles" was suddenly in the spotlight. This drinking club has its roots in the Eighth Air Force in Europe during World War II. The airmen started the club for fun to escape the constant horrors of war. If someone asked a member, "Are you a Turtle," he had to reply with a socially unacceptable version of "You bet your sweet booty I am." If he was not able or willing to respond, he had to buy drinks for everyone listening. Members waited until the worst possible moment to blast one another with the question. A great example came early in Wally Schirra's Mercury flight.

On October 3, 1962, Schirra flew the third US orbital mission aboard Sigma 7. Three and a half minutes after liftoff, CAPCOM Deke Slayton gave the notorious prankster Schirra a dose of his own medicine. As the world listened, Slayton asked, "Are you a Turtle today?"

Thinking quickly, Schirra told Slayton he was switching to his onboard voice recorder. There he gave the correct answer and returned to the normal air-to-ground loop. "You bet," he told Slayton.

Deke responded, "Just trying to catch you on that one." After the flight, Schirra played the voice recording for Walt Williams, Mercury's operations director. All involved were satisfied.

---

[4] You can hear Nafzger's audio by selecting the "Apollo 7 audio as recorded at Corpus Christi" link on the Apollo 7 page of Colin Mackellar's Honeysuckle Creek website: https://www.honeysucklecreek.net/msfn_missions/Apollo_7_mission/index.html.

Six years later, Schirra returned the favor. Slayton wasn't the only target. Schirra drifted away from the camera and Cunningham floated in view, holding a different sign.

> **SWIGERT:** Here comes another one. It says. "Paul Haney, are you a Turtle?"
> **EISELE:** You'll get a gold star; perfect score!

Paul Haney was the Public Affairs Officer providing commentary during the telecast. He was stuck, having no way to answer.

As the crew waited for the replies from Slayton and Haney, they gave the viewing audience a tour of the cabin. Gene Cernan got on the communications loop. "Wally, this is Gene. Deke just called in, and we've got your answer, and we've got it recorded for your return."

Schirra replied, "Roger. Real fine." A few minutes later, Schirra asked about the unfinished business.

> **SCHIRRA:** Have you got Haney's answer yet?
> **SWIGERT:** No, Haney isn't talking, Wally. Somebody tells me he isn't talking, but just buying.
> **SCHIRRA:** He is buying. Thank you very much.

A few minutes after the broadcast concluded, Schirra told Swigert, "Remind Deke it took six years to get that question back to him."

The television broadcast was another hit. The mission was a technical success and certified the Command/Service Module was ready for a lunar flight.

It was full steam ahead for the Moon. The question was if December would be soon enough to beat the Russians. On November 14, the *New York Times* reported the Soviets successfully flew an unmanned Zond 6 around the Moon.[5] The *Times* also reported crew selection for Apollo 10: Commander Tom Stafford, Lunar Module Pilot Gene Cernan, and Command Module Pilot John Young. The headlines stated, "Apollo 10 May Land on the Moon."[6]

---

[5] "Zond 6 Loops Moon," *New York Times*, November 14, 1968, https://timesmachine.nytimes.com.
[6] "Apollo 10 May Land on Moon; Prime Crew Named For '69 Shot," *New York Times*, November 13, 1968, https://timesmachine.nytimes.com.

First live television broadcast from Space, Apollo 7, October 14, 1968.
Wally Schirra on the right holds up a message while Donn Eisele observes.
Photo credits: NASA, Dick Nafzger, Colin Mackellar.

The TEXAS tracking station in 1967, with the main operations building
in the center of photo. Credit: Corpus Christi Caller Times.

10

IN THE BEGINNING, GOD

Apollo is a program to inspire the romantic, challenge the technologist, awaken a nation to its greater capabilities. "No single space project will be more impressive to mankind," President Kennedy said when he initiated the man-to-the-Moon project in 1961, "and none will be so difficult or expensive to accomplish."

Difficult and expensive it has been. From the outset the $24 billion Apollo project has been beset by engineering delays, accelerating costs, political controversy, criticism from scientists and tragedy on the launching pad. And there has always been that haunting prospect that, for all the effort and cost, the Soviet Union might reap the glory by getting its cosmonauts to the Moon first.[1]

APOLLO 8

With these words, John Noble Wilford of the *New York Times* painted the backdrop for the Apollo 8 mission. On Saturday, December 21, 1968, a nervous Mike Collins manned the CAPCOM console as friends and former crewmates were preparing for a daring flight atop the mighty Saturn V.

Collins was their original Command Module Pilot. During handball games earlier in 1968, he noticed something unusual with his left leg.

---

[1] John Noble Wilford, "Space: Step Toward the Moon," *New York Times*, October 13, 1968, https://timesmachine.nytimes.com.

Collins tried to ignore it until his leg started to buckle while walking down stairs. Tingling and numbness spread, and he reluctantly went to see the NASA flight surgeon.[2] "Every pilot knows that if he walks into the doctor's office on flying status, there are only two ways he can walk out: on flying status or grounded," explained Collins. "Since his status could only change for the worse, why risk it? Maybe it will get better… but this one I knew wasn't getting better."[3]

The NASA physician referred him to a specialist who discovered a bony growth between two vertebrae in his neck. It was pushing against his spinal cord, and surgery was the only remedy. On the night before surgery, Collins and his wife Pat ate a quiet dinner at a Mexican restaurant on the Riverwalk in downtown San Antonio. They discussed their five years at NASA and wondered what the future would hold. The date was July 21, 1968, exactly two years after his splashdown with John Young on Gemini X.[4]

The spinal fusion surgery was successful. Collins wore a neck collar for three months while going through physical therapy and receiving positive reports on his progress. Due to the surgery, he was replaced on the prime crew by his backup CMP, Jim Lovell, but he wanted to help with the mission in any way possible. While the crew members spent much of their time in simulators, Collins represented them in numerous planning meetings.[5]

There was great apprehension on the dawn of the Apollo 8 flight, the first manned launch of a Saturn V. At 7:51 a.m., the five first-stage engines roared to life, and the 6.5-million-pound rocket slowly lifted off the pad. To the flight controllers, everything looked normal. But inside the cabin, Bill Anders was surprised at the thrashing they experienced. The Saturn V acted like a giant whip. Its center of mass at launch was located in the first stage, less than 100 feet above the ground. Any vibrations below were increased dramatically at the top of the rocket.[6]

---

[2] Flight surgeons are primary care physicians serving military aviation personnel and astronauts.
[3] Collins, *Carrying the Fire*, 288.
[4] Collins, *Carrying the Fire*, 290.
[5] Collins, *Carrying the Fire*, 298.
[6] NASA, "Apollo/Saturn Postflight Trajectory," October 6, 1969, https://archive.org/details/nasa_techdoc_19920075301/page/n13/mode/2up, 3-18.

To remain on the right trajectory, the engines made constant adjustments directing the thrust, which produced violent jolts in the Command Module. Bill Anders was sure the fins on the base of the launch vehicle were grinding against the launch tower. Anders "tried to find an instrument or a gauge to monitor, something that would confirm the disaster unfolding beneath him, but his head was being shaken with such force that he couldn't focus or think," writes Robert Kurson.[7]

The launch phase proceeded as planned, and twelve minutes later, Apollo 8 was in orbit 100 nautical miles above Earth. After evaluating the spacecraft systems, it was time for man to leave his home planet and head for the Moon. The launch vehicle's third stage, the S-IVB, would reignite for five minutes in a maneuver called "Translunar Injection" (TLI). Two hours and twenty-seven minutes into the mission, Mike Collins told the crew to proceed with the historic burn. "Apollo 8, you are Go for TLI, over."

Collins lamented the official terminology did not capture the historical significance of the moment:

> If you've got a situation where a guy with a radio transmitter in his hand is going to tell the first three human beings they can leave the gravitational field of Earth, what is he going to say? He's going to invoke Christopher Columbus or a primordial reptile coming up out of the swamps onto dry land for the first time, or he's going to go back through the sweep of history and say something very, very meaningful, and instead he says, what? "You're Go for TLI?" I mean, there has to be a better way, don't you think, of saying that? [Laughing] Yet that was our technical jargon.[8]

The burn lasted five minutes and raised the spacecraft's velocity from 25,500 feet per second to 35,000 fps (15,108 to 20,737 mph). This maneuver created a large elliptical orbit around Earth, intersecting the Moon a few days later.

---

[7] Kurson, *Rocket Men*, 152.
[8] Michael Collins, interview by Michelle Kelly for NASA Johnson Space Center Oral History Project, October 8, 1997, 25.

MOON-BOUND

The following day, Mike Collins was back at the CAPCOM console and read the latest version of the news, assembled by Public Affairs Officer (PAO) Paul Haney.

> COLLINS: I've got a Haney special here for you. The Interstellar Times latest edition says the flight to the Moon is occupying prime space on both paper and television. It's *the* news story. The headlines of the Post say, "Moon, here they come." We understand that Bill Anders will be in private conversation or communication today with an old man who wears a red suit and lives at the North Pole.
> ANDERS: Roger. We saw him earlier this morning, and he was heading your way.

Then the conversation turned to football scores. Mike Collins and Frank Borman graduated from the US Military Academy at West Point, (Army), while Bill Anders and Jim Lovell graduated from the US Naval Academy in Annapolis. Collins wanted to rub in the fact his Army Black Knights defeated the Navy Midshipmen in their annual rivalry. He said, "I've got another score for you when you are ready to copy."

Anders told him to proceed.

Collins continued, "Navy 14, Army 21. Would you like for me to repeat that?"

Anders quickly countered, "You are very garbled, Houston. I'm unable to read. Will call you back in another year."

As Apollo 8 journeyed away from home, its velocity gradually decreased. During the outbound trip, the crew photographed Earth, performed navigational sightings, and monitored the spacecraft systems. Ken Mattingly replaced Collins as CAPCOM before the crew put on their first television broadcast.

> PUBLIC AFFAIRS OFFICER (DOUG WARD): This is Apollo Control at 31 hours, 5 minutes into the mission. We are standing by at this time to receive the first television transmission from the spacecraft. Now there is a certain amount of uncertainty just as to when that signal will be received at the station at Goldstone and transmitted to Houston.

The black-and-white image appeared right on time, showing Commander Frank Borman waving at the camera. Borman greeted the audience and explained, "We're rolling around to a good view of the Earth, and as soon as we get to the good view of the Earth, we'll stop and let you look out the window at the scene that we see. Jim Lovell is down in Lower Equipment Bay preparing lunch, and Bill is holding a camera here for us both."

As Anders focused on Lovell in the lower part of the Command Module, the camera rotated, making Lovell appear upside down. Laughter and contorted necks abounded in Mission Control. Mattingly said, "You got everybody standing on their heads down here."

The crew used a wide-angle lens inside the cockpit. Viewers waited for scenes of Earth from deep space which were supposed to highlight the show. The crew had a telephoto lens with a narrow field of view for that purpose. When Anders changed to the telephoto lens, there was no image, to the great disappointment of all involved.

After the transmission, engineers on the ground found a solution, and CAPCOM Mike Collins read procedures to the crew the following day. As the time approached for the second television broadcast of the mission, Borman and Collins strategized about the upcoming broadcast.

> **BORMAN:** We are looking at the Earth right now, and there is a spectacular long, thin band of clouds, looks like it may be a jet stream. It's absolutely spectacular, going almost halfway around the Earth.
> **COLLINS:** Well, you might want to repeat that during the TV narrative, and we'd also like for you, if possible, to go into as much of a detailed description as you poets can on the various colors and sizes of those things and how the Earth appears to you, in as much detail as you possibly can muster.
> **BORMAN:** Roger. I figure we will have to do that because I bet the TV doesn't work.
> **COLLINS:** Well, we won't take that bet, but anyway, we are standing by for a nice lurid description. And we suggest you talk a little bit slower than you did yesterday.

Half an hour later, Mission Control received an image, and Earth emerged in the top right corner. Jim Lovell said, "It's about as big as the end of my thumb." Since there was no monitor for the camera with a telephoto lens, Mike Collins had to direct them as they placed Earth in the middle of the picture. People on Earth could see the entire planet on a live TV broadcast from 180,000 miles.

The western hemisphere was in view. Lovell described the cloud patterns and the vivid colors. Anders added, "I hope everyone enjoys the picture that we are taking of themselves."

> LOVELL: Mike, what I keep imagining is, if I'm some lonely
> traveler from another planet, what would I think about the
> Earth at this altitude, whether I think it'd be inhabited or not.
> COLLINS: You don't see anybody waving. Is that what you are
> saying?
> LOVELL: I was kind of curious whether I would land on the
> blue or the brown part of the Earth.
> ANDERS: You better hope that we land on the blue part.
> COLLINS: So do we, babe.

Borman ended the telecast, "Okay, Earth. This is Apollo 8, signing off for today."

Collins replied, "Good show, Apollo 8. We appreciate it. See you mañana."

Ten minutes later, the spacecraft left Earth's sphere of influence and was now controlled by the lunar gravitational field. The spacecraft slowed to under 4,000 feet per second but would gradually accelerate as it drew close to the Moon.

PAO Paul Haney commented, "Apollo 8 has passed into the Moon's sphere of influence and quite literally, this is a historic landmark in space flight because for the first time, a crew is literally out of this world. They are under the influence of another celestial body, the Moon."

Mission Control changed its charts and reporting procedures for communicating the location and velocity of the spacecraft, moving from Earth-based calculations to Moon-based. The change in reference point made the spacecraft appear to have a new position and velocity. Flight Dynamics Officer (FDO, call sign pronounced "FIDO") Phil Shaffer made the colossal mistake of trying to explain this to the media in a press conference after the shift changeover. In *Carrying the Fire*, Mike Collins tells the story:

Never has the gulf between the non-technical journalist and the non-journalistic technician been more apparent. The harder Phil tried to dispel the notion, the more he convinced some of the reporters that the spacecraft actually would jiggle or jump as it passed into the lunar sphere. Big as a professional football player, red-faced and sweating, Phil delicately re-examined his tidy equations and patiently explained their logic. No sale. Wouldn't the crew feel a bump as they passed the barrier and become alarmed? How could the spacecraft instantaneously go from one point in the sky to another without the crew feeling it? The rest of us smirked and tittered as poor Phil puffed and labored, and thereafter we tried to discuss the lunar sphere of influence with Phil as often as we could, especially when outsiders were present.[9]

FDO David Reed had a more straightforward way of explaining the phenomenon of leaving Earth's sphere of influence. "Phil could have explained that it was like cresting over a steep hill onto the downward side that wasn't quite as steep. You slow down as you approach the crest and pick up speed again as you head down the other side."[10]

HEIDI, ROUND TWO

Collins decided to have a little more fun before turning the CAPCOM duties over to Ken Mattingly. A month earlier, a televised football game between the bitter rivals Oakland Raiders and New York Jets ran long, exceeding its allotted three-hour time slot. NBC televised the well-promoted late-season matchup, and it was must-see TV. The back-and-forth game saw the Jets take a three-point lead with one minute remaining. Unfortunately, the network scheduled the movie *Heidi* to begin at 7:00 p.m. NBC made a previous agreement with the movie's sponsor Timex stating the show would begin on time.

As the hotly-contested game neared its conclusion, *Heidi* fans called in mass, demanding they start the movie at the scheduled time, which clogged the switchboards. Football fans also called to voice their opinions. NBC executives conferred from their homes and decided to remain with the game. They attempted to get that message to the

---

[9] Collins, *Carrying the Fire*, 308–309.
[10] David Reed, email message to author, January 10, 2020.

broadcast operations team, but since the network did not have a special phone line by which executives could contact operations, the message could not get through. At 7:00 p.m. on the east coast, as quarterback Daryle Lamonica was leading the Raiders in a furious last-minute drive, the images of dirty, sweaty behemoths switched to a little girl peacefully walking in the Swiss Alps with her grandfather.

The New York football fans were furious. Thousands called NBC, jamming switchboards and blowing fuses. Many went to bed thinking the Jets won, only to have morning newspapers reveal the Raiders scored two touchdowns in the dying seconds to win the game. To prevent future debacles, NBC installed a hotline for their executives to get through to those in broadcast operations. It became known as the "Heidi Phone."[11]

Collins wanted to get a reaction from Frank Borman, a huge football fan. The previous day's broadcast cut into another televised game, a playoff matchup between the Baltimore Colts and the Minnesota Vikings on the Columbia Broadcasting System (CBS). Fortunately, the Apollo 8 telecast interrupted halftime and just two minutes of action. CBS reported only 2,000 phone calls in protest.[12]

Mike Collins said, "We are having a playback of your TV show and are all enjoying it down here. It was better than yesterday's because it didn't preempt the football game."

Borman was alarmed. "Don't tell me they cut out the football game. Didn't they learn from *Heidi*?"

"Well, you and *Heidi* are running neck-and-neck in the telephone call department."

APPROACHING THE MOON

The famous American sportscaster Howard Cosell gave Raider quarterback Daryle Lamonica the nickname "Mad Bomber" because of his frequent long passes. Completing a long bomb requires great precision to hit a distant moving target with exact placement, velocity, and trajectory. Going into lunar orbit has similar dynamics but is

[11] Tim Hollis, "The Heidi Game in Birmingham," *Birmingham Rewound*, accessed January 15, 2020, https://www.birminghamrewound.com/features/heidi.htm.
[12] "Shades of 'Heidi!' Astronauts Eclipse Title Game on TV," *New York Times*, December 23, 1968, https://timesmachine.nytimes.com.

exponentially more complex. The controllers launched the spacecraft a quarter of a million miles from its target, attempting to slide it 60 nautical miles in front of the leading edge of the Moon, which had traveled over 130,000 miles around Earth since liftoff.

As the crew approached their rendezvous with the Moon, they were oriented backward in its shadow and could not see their target. They had to trust the computations from Mission Control indicating they were neither going to hit the Moon nor be too far away from it. The former would cause sudden death in the non-football sense, and the latter would miss the free return trajectory back to Earth in case the Service Module engine failed.

If the numbers were correct, Apollo 8 would go around the western limb of the Moon and lose radio contact with Houston at 68 hours, 58 minutes, and 4 seconds into the mission. Ten minutes later, they would burn their Service Module engine against the direction of travel for four minutes, slowing its velocity down to the correct speed for lunar orbit.

The large engine bell at the back of the Service Module led the way while the astronauts' heads were toward the lunar surface. A clock in the front of the Mission Control Center counted down the minutes and seconds until Loss of Signal (LOS), when the Moon would block the radio waves. One minute before LOS, the crew turned on the cabin tape recorder.

CAPCOM (JERRY CARR): Safe journey, guys.
ANDERS: Thanks a lot, troops.
LOVELL: We'll see you on the other side.

At the exact second, the three astronauts lost radio signals from Houston becoming the first humans in history to be out of communication with Earth. They were amazed at the precision. Borman joked, "That was great, wasn't it? I wonder if they turned the radio off."

Anders continued, "Chris Kraft probably said, 'No matter what happens, turn it off.'"

Apollo 8 traveled around the back side of the Moon. The crew had ten minutes to prepare for the Lunar Orbit Insertion (LOI) burn, going through their detailed checklist. It would be eerily dark until three minutes before the LOI maneuver. They could not see the Moon, but they knew it was there.

As the ignition time approached, Borman, Lovell, and Anders saw the Sun's rays begin to highlight the lunar surface below. Commander Borman reminded everyone to keep their mind on the business at hand. They would have plenty of time to look at the Moon later.

If the Service Module engine did not ignite, they would remain on their free return trajectory, slingshot around the Moon, and return to Earth. If the burn were long, too much speed would be scrubbed, and they would crash into the lunar surface. If the burn were short, they would maintain too much velocity to be captured by the Moon's gravity and fly off into space. It had to be perfect.

The engine ignited right on time, and the crew closely monitored the change in velocity, pressure in propellant tanks, and spacecraft attitude. The astronauts carefully watched the burn, ready to shut off the engine manually in case it fired too long. The engine stopped right when it achieved the desired change in velocity. Everything was "nominal," according to plan, and they spent the next few minutes going through the post-burn checklist. The Russians had accomplished many firsts in the space race, but now three Americans were the first in lunar orbit.

Houston was in the dark regarding what was transpiring behind the Moon. It was 4:00 a.m. December 24 in Houston, and Susan Borman, Marilyn Lovell, and Valerie Anders were listening intently to the squawk boxes in their homes.[13] Public Affairs Officer John McLeaish anxiously kept the world informed:

Now we are in our period of longest wait thus far in the mission. We are 19 minutes, 50 seconds away from acquisition at this time. During Mission Control simulations, this was a good time for coffee breaks for the flight controllers. But that is not true today. Continuing to monitor, this is Apollo Control, Houston.

Two clocks at the front of Mission Control counted down to Acquisition of Signal (AOS). The top clock was set for the no-burn

[13] Squawk boxes were small speakers allowing families to hear conversations between astronauts and Mission Control.

scenario where the free return trajectory would send them back to Earth. The lower was set to the moment they would acquire communication if all went nominally. There was nothing Glynn Lunney's team of controllers could do but wait.

PAO John McLeaish kept the world updated through his commentary on a communications loop called the Public Affairs Release, or PAO Release. It was the source from which the media and everyone in the world except for the flight control team received their air-to-ground communications. PAO Doug Ward explains, "It was configured that way, in part, because if there was an emergency and we chose to interrupt the communications, keying our microphone would cut out the air-to-ground."[14]

LUNAR ORBIT

The AOS clock for the no-burn situation hit zero without a signal from the spacecraft. Ten minutes later, Houston acquired signals from the Command Module right on schedule. Voice contact soon followed. John McLeaish could not hide his excitement.

> We've got it. We've got it. Apollo 8 now in lunar orbit. There is a cheer in this room. This is Apollo Control, Houston, switching now to the voice of Jim Lovell.

Lovell reported their burn went well, and the resulting orbit was 169 by 60.5 nautical miles. CAPCOM Jerry Carr responded, "Good to hear your voice." After 35 minutes of helplessness, the flight controllers could now study the spacecraft data.

Carr asked, "What does the ol' Moon look like from 60 miles? Lovell replied, "The Moon is essentially gray, no color. Looks like plaster of Paris, or sort of a grayish beach sand. We can see quite a bit of detail."

Lovell continued the commentary of the lunar terrain while Anders took photographs and Borman flew the spacecraft.

The crew photographed several predetermined targets of opportunity on the near and far side, as well as potential Apollo landing

---

[14] Doug Ward, interview by author, March 16, 2019.

sites. They began their second nearside pass with a television broadcast showing craters out their window. They named some in honor of their fellow astronauts who died in the Apollo 1 fire, Grissom, White, and Chaffee. They also honored those who perished in various crashes, Ted Freeman, Elliott See, Charlie Bassett, and C.C. Williams.

The crew named craters for themselves, NASA leaders, and even CAPCOMs Jerry Carr and Mike Collins. John Aaron, a flight controller monitoring the electrical and environmental systems, was "bowled over" when they named a crater after him in appreciation for his help with some issues earlier in the flight.[15] "If he'll keep looking at the systems," Anders added in jest.

Aaron, who would become a Control Center legend with monumental decisions made during Apollo 12 and Apollo 13, looked up in stunned disbelief at hearing his name called from lunar orbit. Carr watched him and notified Anders, "He just quit looking."

Phil Shaffer and the other Flight Dynamics Officers noticed something was interfering with the spacecraft's orbits, causing their predicted trajectories to be incorrect. They were dealing with varying concentrations of mass, called mascons. "They are areas of much denser rock than there is in other places," said Shaffer. "We don't know whether they came from outside or whether they're necessarily deep lava pools that have hardened or exactly what they are, but the Moon is a very non-homogeneous body."[16]

With the spacecraft still facing backward, the crew performed an 11-second burn which circularized their orbit about 61 nautical miles above the lunar surface. When an orbiting spacecraft fires its engine, it will return to the same point of the firing on the next orbit. But the opposite side of the orbit is affected by the burn. If the burn is in the direction of travel, the acceleration will increase the altitude on the opposite side. If the burn is against the direction of travel, the reduced velocity will lower the altitude on the opposite side.

An example of the former is the Translunar Injection which adjusts the circular Earth orbit into a highly elliptical orbit intersecting the

---

[15] John W. Aaron, interview by Kevin M. Rusnak for NASA Johnson Space Center Oral History Project, January 18, 2000, 24.
[16] Philip C. Shaffer, interview by Carol Butler for NASA Johnson Space Center Oral History Project, January 25, 2000, 63.

Moon's sphere of influence. An example of the latter is the Apollo 8 circularization burn which occurred at the low point of an elliptical orbit and reduced the altitude on the other side of the Moon to match it.

As the crew came around the east limb of the Moon on their third orbit, they reported the details of the successful burn. Then they quietly did their tasks as Houston monitored their spacecraft. Mike Collins replaced Jerry Carr as CAPCOM, and Borman asked for the start and stop times for the primetime television transmission occurring on the ninth orbit. Borman mentioned he wanted it to include their crossing the terminator, where the sunlight ends and darkness begins, producing long and dramatic shadows. This was in stark contrast to the views when the Sun was directly overhead, washing out the shadows. Collins said he would give those details on their next frontside pass.

ROD ROSE

Bill Anders reported on the targets of opportunity he had photographed so far. Then Frank Borman asked if Rod Rose was nearby. Rose was Chris Kraft's technical assistant in the Flight Operations department and was involved in planning the flight. He was also a neighbor of the Borman family and, like Frank, was a leader at St. Christopher's Episcopal Church. The church scheduled Borman to be a lay reader at the Christmas Eve Communion Service. He and Rose learned about the assignment in a meeting with the minister, but the details of the Apollo 8 mission were still secret. All Borman could tell the minister was, "I'm going to be on travel."

Rose later explained, "I knew where he was going, so I took Frank outside, and I said, 'Frank, I think we can work this. If you read the prayer from the Moon, I'll get it taped in the MCC and whip the tape over to the church, and we can play it in the service.'"[17]

> COLLINS: Rod Rose is sitting up in the viewing room. He can hear what you say.
> BORMAN: I wonder if he is ready for experiment P1?
> COLLINS: He says thumbs-up on P1.

---

[17] Rodney G. Rose, interview by Kevin M. Rusnak for NASA Johnson Space Center Oral History Project, November 8, 1999, 29.

BORMAN: Okay. This is to Rod Rose and the people at St. Christopher's, actually to people everywhere: Give us, O God, the vision which can see Thy love in the world in spite of human failure. Give us the faith to trust Thy goodness in spite of our ignorance and weakness. Give us the knowledge that we may continue to pray with understanding hearts. And show us what each one of us can do to set forth the coming of the day of universal peace. Amen.
COLLINS: Amen.
BORMAN: I was supposed to lay-read tonight, and I couldn't quite make it.
COLLINS: Roger. I think they'll understand.

A few minutes later, Collins informed the crew they had two and a half minutes before LOS, and all systems looked good.

COLLINS: We'll have some more information on the TV on the next rev. We're not planning any big change in the time, just extend them a little bit, I think, close to the terminator.
BORMAN: I'd like to get the terminator if we could, and we've got a little message, and that's it.
COLLINS: Roger. We'll do that the next time you come around.

On the following near side pass, Anders asked for the latest news from Earth. Collins said, "San Diego welcomed home today the Pueblo crew in a big ceremony. They had a pretty rough time of it in the Korean prison.[18] Christmas cease-fire is in effect in Vietnam, with only sporadic outbreaks of fighting. And if you haven't done your Christmas shopping by now, you better forget it."

Commander Borman was concerned about the Trans-Earth Injection scheduled for later in the day, the burn which would send the spacecraft back home. It had to be perfect. He had a few hours of sleep earlier in the day, but Lovell and Anders had not. He wanted them to be fresh for the critical maneuver. On the seventh orbit, Borman alerted Collins he was suspending their duties for a few hours so they rest.

---

[18] The *USS Pueblo* was a Navy intelligence vessel captured by North Korea on January 23, 1968. Over 80 crew members were severely treated, and tortured on a regular basis. On December 23, negotiators reached a settlement that secured the release of the men.

"We're getting too tired." Ten minutes later, Borman reported Lovell was already snoring. Collins replied, "Yeah, we can hear him down here."

CHRISTMAS EVE

While the crew was on the back side of the Moon, there was a personnel change in Control Center. Several teams of controllers rotated in eight-hour shifts throughout the mission. Each team was led by a Flight Director who selected color designations for his troops. Chris Kraft was the first Flight Director in the Mercury program and selected Red. When the final Mercury flight was too long for one shift, John Hodge became the second Flight Director and chose Blue for his team. The ever-patriotic Gene Kranz was the third Flight Director and selected White.

At this point in the flight, Milt Windler's Maroon team took over in Mission Control, and Ken Mattingly replaced Mike Collins at the CAPCOM position. Borman was the only active member of the crew on their eighth frontside pass. Even with the crucial Trans-Earth Injection (TEI) burn looming, he took a few moments for a Christmas Eve chat with Mattingly.

> BORMAN: How's the weather down there, Ken?
> MATTINGLY: It's really beautiful. Loud and clear, and just right in temperature.
> BORMAN: How about the recovery area?
> MATTINGLY: That's looking really good.
> BORMAN: Very good.
> MATTINGLY: They tell us that there is a beautiful Moon out there.
> BORMAN: I was just saying that there's a beautiful Earth out there.
> MATTINGLY: It depends on your point of view.

The clock read 7:30 p.m. in Houston on December 24, an hour before the highly anticipated live television broadcast.

> BORMAN: Hey, Ken, how did you pull duty on Christmas Eve? It happens to bachelors every time, doesn't it?
> MATTINGLY: I wouldn't be anywhere else tonight.

Apollo 8 came around the corner for their ninth orbit while the entire planet waited. An estimated one billion people in 64 countries were tuning in on television or radio. The audience was one-fourth of the planet's population and the most people to have ever listened to a human voice live. At 8:30 p.m., all three major American networks cut into their Christmas Eve programming.

As the world looked at a grainy image, Frank Borman welcomed them. "This is Apollo 8 coming to you live from the Moon." The camera looked out the window at Earth and the lunar horizon. Then they changed windows and pointed the camera at craters passing underneath. The crew described their observations and expressed their thoughts for the next 20 minutes. And then it was time.

For weeks, the astronauts had known they would have an opportunity to give a special message to the people on Earth during their broadcast but did not disclose what they would say. Their families did not know, and their bosses at NASA did not know.

As they approached the terminator and the shadows lengthened across the craters, Bill Anders began:

ANDERS: For all the people back on Earth, the crew of Apollo 8 has a message that we would like to send to you:

In the beginning, God created the heaven and the Earth. And the Earth was without form and void, and darkness was upon the face of the deep. And the Spirit of God moved upon the face of the waters. And God said, "Let there be light," and there was light. And God saw the light, that it was good, and divided the light from the darkness.

LOVELL: And God called the light day, and the darkness He called night. And the evening and the morning were the first day. And God said, "Let there be a firmament in the midst of the waters. And let it divide the waters from the waters." And God made the firmament and divided the waters which were under the firmament from the waters which were above the firmament. And it was so. And God called the firmament heaven. And the evening and the morning were the second day.

BORMAN: And God said, "Let the waters under the heavens be gathered together into one place. And let the dry land appear." And it was so. And God called the dry land Earth. And the gathering together of the waters called the seas. And God saw that it was good.

And from the crew of Apollo 8, we close with good night, good luck, a Merry Christmas, and God bless all of you, all of you on the good Earth.

STUNNED RESPONSE

Anders turned off the camera. Screens went to black as people around the world reflected on the experience. Flight controllers were blown away and simply looked at each other.[19] Robert Kurson captured the sentiments:

> Inside Mission Control, no one moved. Then, one after another, these scientists and engineers in Houston began to cry. The agency had allowed the crew to choose what to say to the world on Christmas Eve—no oversight, no committees, not even a quick glance on the day before they'd departed. It had come as a complete surprise to them. In his studio at CBS, Walter Cronkite fought back tears as he came back on the air.... Watching in Houston, Susan Borman wept. Marilyn Lovell gathered up her kids and they walked, not drove, past the holiday lights in Timber Cove, slow enough to remember them all. Valerie Anders told her children, "That was for the whole world." Across much of the globe, people streamed outside and looked up, trying to pick out the three men who'd just spoken to them, knowing it was impossible, but trying all the same.[20]

Onboard, Anders confirmed the camera was off. Frank Borman was pleased with the technical aspects of the broadcast.

---

[19] Glenn S. Lunney, *Highways into Space: A first-hand account of the beginnings of the human space program,* Self-published, 2014, 150.
[20] Kurson, *Rocket Men,* 263.

BORMAN: Hey, how can you beat that? We just went into the terminator right on time.
LOVELL: Okay, let's get the spacecraft back in even keel again.
BORMAN: All right, let's get the flight plan out here.

The astronauts had no idea of the impact of their words. Mattingly confirmed everything was "loud and clear. Thank you for a very good show." Gene Kranz was not working the mission but was in the Control Center as an observer. "I was enraptured, transported by the crew's voices, finding new meaning in the words from Genesis. For those moments, I felt the presence of creation and the Creator. Tears were on my cheeks."[21]

Decades later, Glynn Lunney reflected on the Genesis reading:

> I feel very emotional and personal about it all. And the choice of that passage could not have been more perfect. I mean, however long it's been, some 30 years later, I'm not sure I would have suggested anything any better than that in retrospect. America has this inventory of people and here come the people when you need them; out they come. And then, to cap it off with that kind of appropriateness and dignity... America asked for them and they got them, and they were the right people and then they did the right thing. They conducted themselves in exactly the right way. I'm sure everybody in America just loved it. It was what they wanted to hear, even if they didn't know it at the time.[22]

Mike Collins added, "Borman, Lovell, and Anders deserved to make it home for that reason alone, for having thought to bring the rest of us to their Moon in humility and reverence. It was a graceful touch."[23]

John Aaron said, "It particularly impacted the mission controllers, because they were sitting there doing this highly technical jargon-oriented job, and then the transition from that to a reading from Genesis, that was the big surprise."[24]

---

[21] Kranz, *Failure Is Not an Option*, 246.
[22] Glynn Lunney, interview by Roy Neal for NASA Johnson Space Center Oral History Project, March 9, 1998, 54–55.
[23] Collins, *Carrying the Fire*, 310.
[24] John Aaron, interview by Kevin Rusnak, January 18, 2000, 25.

Rod Rose took reel-to-reel tapes of Borman's prayer and the Genesis reading to St. Christopher's Episcopal Church. At the same time, Borman, Lovell, and Anders prepared for the all-important Trans-Earth Injection which would send them out of lunar orbit and back to Earth. It would take place a few minutes after midnight Houston time on the far side of the Moon. If all went well, they would emerge around the corner of the Moon at 12:19 a.m.

Public Affairs Officer Doug Ward described the scene in Mission Control as Apollo 8 made its final pass on the near side of the Moon.

> *At the present time, the mood here in Mission Control Center could best be described, I believe, as one of relaxed confidence. Flight controllers are continuing to go over their displays, looking at the systems, getting last minute look at all systems before we lose contact with the spacecraft. And we are again, going back over the figures that have been passed up to the crew, verifying every aspect of this maneuver.*

Ward walked down to the front row of Mission Control and spoke to RETROFIRE Officer Chuck Deiterich, who gave him a copy of the Pre-Advisory Data (PAD) which was going to be read up to the crew by CAPCOM.[25] The PADs were long lists of numbers giving details about upcoming burns, such as the weight of the spacecraft, time of ignition, change in velocity, and length of the burn. It was helpful to the PAO commentators, who could translate the data into meaningful information for the outside world. Doug Ward recalls:

> They wrote the PADs on pressure-sensitive paper, chemically sensitive, and when you wrote on the original, it would make three or four copies if you put enough pressure on it. It would be kind of a pale blue, almost like a blueprint. We were not on the distribution list for those. If we got one, we would have to go down and ask the Flight Dynamics Officer for it. Sometimes they would just write it out for us. But if we did not have that advanced copy, we would just copy the numbers

---

[25] Doug Ward, interview by author, March 16, 2019.

down as they read them and reiterate them when we did the commentary on it.[26]

HOMEWARD-BOUND

The crew carefully progressed through the pre-burn checklists and ensured their guidance and navigation systems were accurate. Jim Lovell performed star sightings to confirm correct alignment for the burn. On the ground, Milt Windler's Maroon team checked and double-checked their numbers. Controllers took one final look at their displays before losing contact with Apollo 8 at 11:42 p.m., December 24.

> *PAO (WARD): At 88 hours and 51 minutes, we show Loss of Signal with the spacecraft. Our next communication with Apollo 8 should come in about 37 minutes. We are now about 28 minutes prior to our Trans-Earth Injection maneuver. As the spacecraft went over the horizon, Capsule Communicator Ken Mattingly passed along for the second time the word that all systems are Go. And we got a very terse "Roger" back from the spacecraft.*

Onboard, Frank Borman said, "It's been a pretty fantastic week, hasn't it?"

Lovell replied, "It's going to get better."

On Earth, everyone held their breath as they waited to learn if the Trans-Earth Injection was successful. Twenty-eight minutes later, Ward gave another update:

> *We are now less than 30 seconds to the scheduled time of ignition for the maneuver to start Apollo 8 on its course back to Earth. In the last 15 seconds prior to ignition, the crew will be burning their Reaction Control System engines to settle propellants. And here in Mission Control Center, we have just counted down to the burn. We should have ignition at this time. That will be a three-minute and eighteen-second burn nominally. It will increase the spacecraft velocity by about 3,522 feet per second or some 2,395 miles per hour. Following the maneuver, the spacecraft should have a*

---

[26] Doug Ward, interview by author, March 16, 2019.

velocity of about 8,800 feet per second, some 6,000 miles per hour. And here in Mission Control, it is relatively quiet, as it has been since we lost communications with the spacecraft as they went over the Moon's horizon. At this point, flight controllers here in Mission Control, as with the rest of the world, they are waiting.

The tension was enormous. NASA management knew the results of the upcoming burn would either open the gate for a lunar landing or put the manned spaceflight program in danger. Soon, flight controllers would see if their careful calculations held up to the ultimate test. The wives and families would know if they would see their loved ones again. Everyone held their breath.

Right on time, at 12:19 a.m., the tracking station at Honeysuckle Creek in Canberra, Australia, detected a radio signal from Apollo 8. The flight controllers cheered. Voice communication would soon follow. But it didn't.

Ken Mattingly called, "Apollo 8, Houston." There was no reply. Again, "Apollo 8, Houston." Again, no reply. Mattingly repeated the call several more times. Finally, at 12:25 a.m., the voice of Jim Lovell pierced through the silence.[27]

> LOVELL: Houston, Apollo 8, over.
> MATTINGLY: Hello, Apollo 8. Loud and clear.
> LOVELL: Roger. Please be informed there is a Santa Claus!
> MATTINGLY: You're the best ones to know.

Their wives and families were ecstatic. Mission Control exulted, and the crew was on the way home. Twenty minutes later, their boss Deke Slayton made a special appearance on the communications loop.

> SLAYTON: Good morning, Apollo 8. Deke here. I'd just like to wish you all a very Merry Christmas on behalf of everyone in the Control Center, and I'm sure everyone around the world. None of us ever expect to have a better Christmas present than this one. Hope you get a good night's sleep from here on and enjoy your Christmas dinner

---

[27] An antenna switch in the wrong position prevented Lovell from being heard earlier.

tomorrow and look forward to seeing you in Hawaii on the 28th.

BORMAN: Okay, leader. We'll see you there. That was a very, very nice ride, that last one. This engine is as smooth as glass.

SLAYTON: Yeah, we gathered that. Outstanding job all the way around.

BORMAN: Thank everybody on the ground for us. It's pretty clear we wouldn't be anywhere if we didn't have them doing it or helping us.

SLAYTON: We concur that.

LOVELL: I concur too.

ANDERS: Even Mr. Kraft does something right once in a while.

SLAYTON: He got tired of waiting for you to talk and went home.

Mattingly turned over the CAPCOM role to fellow astronaut Harrison "Jack" Schmitt. Schmitt, a geologist who would walk on the Moon during Apollo 17, worked extensively with Jim Lovell on lunar landmarks and photography leading up to the mission.

SCHMITT: Typhoid Jack here, and we've got some good words that originated at the Cape with a bunch of friends of yours. And it's sort of in a paraphrase of a poem that you probably are familiar with:

'Twas the night before Christmas and way out in space, the Apollo 8 crew had just won the Moon race. The headsets were hung by the consoles with care, in hopes that Chris Kraft soon would be there.

Frank Borman was nestled all snug in his bed, while visions of REFSMMATs danced in his head;[28] and Jim Lovell, in his couch, and Anders, in the bay, were racking their brains over a computer display.

---

[28] REFSMMAT stands for "Reference to Stable Member Matrix," a set of numbers the computer used to determine the spacecraft's orientation in relation to known reference points.

When out of the DSKY, there arose such a clatter,[29]
Frank sprang from his bed to see what was the matter.
Away to the sextant he flew like a flash,
to make sure they weren't going to crash.

The light on the breast of the Moon's jagged crust,
gave a luster of green cheese to the gray lunar dust.
When what to his wondering eyes should appear,
but a Burma Shave sign saying "Kilroy was here."[30]
[much laughter in Mission Control]

But Frank was no fool; he knew pretty quick
that they had been first; this must be a trick.
More rapid than rockets, his curses they came.
He turned to his crewmen and called them a name.

Now Lovell, now Anders, now don't think I'd fall,
for that old joke you've written up on the wall.
They spoke not a word, but grinning like elves,
and laughed at their joke in spite of themselves.

Frank sprang to his couch, to the ship gave a thrust,
and away they all flew past the gray lunar dust.
But we heard them explain
ere they flew around the Moon,
Merry Christmas to Earth;
we'll be back there real soon.

Great job, gang.

BORMAN: Thank you very much. That's a very good poem. But
in order to win the race, you've got to end up on the carriers.

Borman, Lovell, and Anders won the race a few days later when
they splashed down in the middle of the Pacific Ocean before sunrise.
They landed within three miles of the *USS Yorktown*, smacking the
water with a bone-jarring impact. The spacecraft flipped over with its

---

[29] DSKY (rhymes with "risky") the guidance computer display/keyboard acronym.
[30] Burma Shave was a shaving cream company famous for advertising with rhyming
statements on a series of road signs. "Kilroy Was Here" was a popular meme of a bald
man with a long nose peering over a wall, often seen in graffiti wherever American
soldiers ventured in WWII.

nose in the water. The crew was hanging by their straps in the same position Lovell simulated with Charlie Duke and Stu Roosa. Borman inflated three balloons to right the spacecraft, then vomited all over the cabin. Lovell and Anders, both Naval Academy graduates, were merciless. "Typical Army guy, can't handle the water!"[31]

Once the recovery helicopters delivered them safely onto the aircraft carrier, the raucous celebrations began in Mission Control. A giant American flag flew in front of the main wall map at the front of the room, along with miniature flags at each flight control console. NASA officials and flight controllers packed the room. Boxes of congratulatory cigars were tossed around.

> *PAO (PAUL HANEY): I'm not sure how well our voice is getting out. There is tremendous roar, an undercurrent and roar in the background. I have never seen a degree of this emotional outpouring in any previous mission, including Alan Shepard's. I guess one of the big differences there between that one and this one is Alan is here, standing right in the middle of this one puffing on a long black cigar. I've seen rallies in locker rooms after championship games, happy politicians after elections, but none of them do justice to the spirit pervading this room. I think that many of you can see this tumult. Someone suggested that we have set the American Cancer Society's antismoking campaign back several light years.*

Colonel Frank Borman led the crew off the helicopter and spoke a few words to the *Yorktown* sailors, thanking them for their role in the recovery which meant being away from home on Christmas. Then the crew went downstairs to sickbay, where physicians evaluated them thoroughly for hours.

A *Yorktown* flight surgeon named George Wagoner was invited to be part of the medical examination team. Afterward, a NASA official recognized him and invited Wagoner to join him in the Admiral's dining room to watch a movie. He walked into the viewing area occupied by the three astronauts and only a few others. First, they watched film from the Apollo 8 launch the previous week. Then they watched the first

---

[31] Kurson, *Rocket Men*, 304.

colored movie of the Moon's far side, which had just been developed on *Yorktown*. Wagoner could not believe he was a part of the special viewing.[32]

EARTHRISE

Seven priceless Hasselblad film magazines from the mission arrived in Houston, and George Abbey went to the Control Center photo lab to observe the processing.[33] George Low tasked him to see if there were any extraordinary images after each flight. He joined the lab's chief John Brinkman and a lab technician who developed the film and printed the photos.[34] Amid dramatic images of the Moon's surface, they noticed a remarkable sequence showing Earth rising over the lunar horizon. They were the first to see the historic photographs, taken at 10:38 a.m. on Christmas Eve.

At that time, the nose of the spacecraft pointed toward the Moon's surface with the windows facing forward, allowing Lovell and Anders to see the approaching terrain and better perform their duties. Anders was busy following the photography list on the flight plan when he noticed something coming over the horizon. "Look at that picture over there! Here's the Earth coming up. Wow, is that pretty!"

Borman joked, "Hey, don't take that, it's not scheduled." They scrambled while trying to capture the remarkable scene. As Earth moved from one window to the next, Anders grabbed a Hasselblad camera with color film. The resulting "Earthrise" portrait is one of the most famous in history. Those images were a complete surprise to Abbey. The idea of photographing Earth over the lunar horizon never came to mind in any of the pre-flight preparations.

---

[32] George Waggoner, "Apollo 8 Recovered by Yorktown," Yorktown Stories, accessed January 15, 2020, Video, 4:10, https://www.ussyorktown.org/yorktown-stories/apollo-8-recovered-by-yorktown/.

[33] NASA began using cameras from the Swedish company Hasselblad in 1962. Wally Schirra owned an iconic Hasselblad 500C and recommended the medium-format camera for his Sigma 7 mission. NASA was thrilled with the results. The modular system allowed for interchangeable lenses on the front and film magazines on the back. The 2.25-inch square negatives and high-quality provided crisp images when significantly enlarged. Modified Hasselblad 500EL cameras were used for Apollo missions. *See* https://www.hasselblad.com/about/history/hasselblad-in-space/ accessed July 13, 2020.

[34] Cassutt, The Astronaut Maker, 110.

THE CLIMAX

For many in NASA, Apollo 8 was the highlight of their careers. Ken Mattingly was involved with Apollo 11, Apollo 13, and Apollo 16, but he said, "Being part of Apollo 8 made everything else anticlimactic. Our purpose was to go land on the Moon, but somehow the participation, the angst that went with that Apollo 8 mission was far more electrifying."[35]

Borman concludes the Genesis reading as Apollo 8 approaches darkness. The crater Sinas, in the middle of the Sea of Tranquility, is near the right edge of the picture. Screen capture of NASA TV.

Earthrise photographed from Apollo 8, Christmas Eve, 1968. Photo credit: NASA AS08-14-2383.

---

[35] Ken Mattingly, interview by Rebecca Wright, November 6, 2001, 21.

11

STROKING TEST

SPIDER AND GUMDROP

President Kennedy's deadline loomed on the horizon, forcing a rapid sequence of flights. "We were receiving the documents for the next mission before the previous one had flown, so it was very, very hectic, but very satisfying," said Mike Dinn, Deputy Director of the Honeysuckle Creek tracking station in Australia. He added, "It was satisfying at the station level that our opinions and our views and our contributions were seriously listened to, and more often than not, acted upon, which was refreshing."[1]

Less than 10 weeks after Apollo 8 splashed down, it was time for the next mission. The intricate Apollo 9 was "a test pilot's feast," said Apollo historian Andrew Chaikin. "In truth it was far more difficult, more ambitious, and in some ways more dangerous than Apollo 8."[2]

Deke Slayton selected Jim McDivitt to command the Lunar Module's maiden voyage. Many within NASA thought McDivitt or Frank Borman would be the first astronaut to attempt a lunar landing. McDivitt, a veteran of 145 combat missions in the Korean War, excelled at every assignment as an engineer, pilot, and leader. Highly respected by NASA officials, he would later be chosen to manage the Apollo Spacecraft Program Office after George Low became NASA's Deputy

---

[1] Mike Dinn, interview by Colin Mackellar for honeysucklecreek.net, accessed July 20, 2021, Audio, 13:32,
https://honeysucklecreek.net/audio/interviews/Mike_Dinn_up_to_A11.mp3.
[2] Andrew Chaikin, *A Man on the Moon*, 141.

Administrator.[3] McDivitt and his crewmates Dave Scott and Rusty Schweickart trained together since being selected in 1966 as the backup crew for the first manned Apollo mission.

The three men came to NASA to test and fly, not to be celebrities. Being overshadowed by high-profile missions before and after did not take away from their excitement of flying the first manned Lunar Module (LM). Test pilots lived for being the first to break in new machines. With no heat shield, the LM was truly a spaceship, only capable of operating beyond Earth's atmosphere.

To call Apollo 9 complex would be an understatement. It was a low Earth orbit simulation of upcoming lunar missions. The crew, flight controllers, and mission planners were stretched to their limits. The two-stage Lunar Module had to be certified for manned lunar flights, which necessitated tests of its descent engine, ascent engine, guidance, communication, environmental control, electrical power systems, and ability to rendezvous and dock with the Command/Service Module (CSM). Though it proved worthy on the previous missions, the CSM had to be tested for rigorous tasks required of it during future lunar flights, plus unexpected scenarios. In addition, the spacesuit and Portable Life Support System (PLSS) backpack had to be evaluated in the harsh environment of space.

Apollo 9 was demanding for the crew and Mission Control, who had to master many of the procedures involved with lunar flights. The mission being flown in Earth orbit complicated everything since there were gaps in the tracking station coverage. All tests had a strict timeline, requiring the spacecraft to be in view of tracking stations so the ground could record the necessary data. On lunar flights, unless on the far side of the Moon, the spacecraft was continuously tracked by the Deep Space Network, which had 210-foot dishes strategically placed in Madrid, Spain; Canberra, Australia; and Goldstone, California. Flight Dynamics Officer David Reed worked closely with the Apollo 9 astronauts as they developed the procedures for the flight. Reed says, "David Scott and I fully agree that it was THE most complex and difficult mission ever flown in Apollo."[4]

---

[3] NASA's Deputy Administrator is the second in command of the space agency.
[4] David Reed, email message to author, January 11, 2020.

STROKING TEST

Apollo 9 launched successfully on March 3, 1969. After entering Earth orbit, Dave Scott piloted the CSM away from the S-IVB booster, returned, and docked with the Lunar Module. The CSM extracted the LM from the booster as it would on future lunar flights. They fired the Service Module's Service Propulsion System (SPS) engine six hours into the flight to test the CSM digital autopilot and demonstrate the ability of one engine to control two docked spacecraft. After they moved a safe distance away from the empty third stage, the S-IVB fired again to simulate a Translunar Injection. The crew filmed the S-IVB burn, the only time such a maneuver was viewed from close range.[5]

> CAPCOM (STU ROOSA): We show ignition on the S-IVB.
> SCHWEICKART: It's on the way.
> McDIVITT: Just like a bright star disappearing into the distance.
> ROOSA: Is there quite a bit of debris kicked out there, Apollo 9?
> McDIVITT: Looked like a real clean burn.
> SCHWEICKART: Yeah, you could see a lot of stuff coming out when he just started up, but then it just went into a nice bright light.
> ROOSA: Beautiful!

The burn lasted for one minute.

> ROOSA: And the S-IVB has shut down, Apollo 9.
> McDIVITT: Roger. He's just a speck in the distance right now.
> ROOSA: Okay. Now that we've got him out of the way, back with the business at hand.

On Day 2, the crew further evaluated the spacecraft in a docked configuration. Specifically, how stable would the docked CSM/LM airframe remain during the burst of a gimballing engine? The SPS engine was attached to the Service Module by a gimbal ring so the engine could pivot, thereby controlling the spacecraft's attitude and

---

[5] The 2019 Apollo 11 Documentary directed by Todd Douglas Miller used this footage for the Translunar Injection scene.

direction of travel. Engineers developed a "Stroking Test" that would introduce oscillations along the pitch axis (up and down) during a burn. These oscillations were programmed into the computer with a maximum amplitude (range) of 1.1 degrees and lasted for seven seconds. During that time, the structural bending of the CSM/LM airframe was measured.[6]

The test took place in two parts. The first was performed at only 40 percent amplitude to ensure the oscillations would not damage the spacecraft. During the 14th orbit, Flight Director Pete Frank told CAPCOM Stu Roosa to communicate the crew was Go for the burn. As they passed over Texas, McDivitt ignited the SPS engine for 110 seconds, including the seven seconds of programmed oscillations. Mission Control evaluated the results, which showed the stresses on the airframe were lower than expected. It was safe to proceed to the second phase with full amplitude oscillations, scheduled to occur a few hours later.

Between the SPS burns, the crew had time to talk about some "little interruptions" they had the previous night. Several times they heard air traffic controller communications with airplanes. They mentioned the conversations earlier, but CAPCOM Roosa thought they were joking.

As Houston tried to find the source of the transmissions, Jim McDivitt said, "I'll give you a clue. They've got a runway that is 112, and they have a taxiway 112. They fly a whole bunch of different kinds of airplanes: Mohawks, and C-47s, and 01s."

Dave Scott said, "And if you really wanted, you could call Green Hornet 35 or Blackhawk 15." Scott was impressed at how well the air traffic controllers performed. "Chris Kraft ought to incorporate these guys into the communications network."

Schweickart said, "Actually, it was one of the better tower operators I've heard. The guy really had a lot of traffic and he was doing pretty good."

Roosa replied, "I guess it's alright as long as you don't have to get clearance through that tower."

The second part of the stroking test took place at the end of the 16th revolution. The crew restarted the SPS engine and waited 60 seconds

6 NASA, "Apollo 9 Mission Report," 8-25, 8-26,

allowing it to stabilize. Then the computer excited the gimbals producing oscillations for seven seconds. After the pre-programmed oscillation period, the engine ran normally for another three and a half minutes. Again, the effects of the stroking inputs were lower than expected, giving confidence the spacecraft could handle oscillations without sustaining damage.[7] This test would play an important role in the success of Apollo 16 three years later.

On Day 3, McDivitt and Schweickart entered the Lunar Module *Spider* through the docking tunnel while Scott remained in the Command Module *Gumdrop*. They powered up the LM, checked the systems, and performed a burn of the descent engine while still docked with the CM. The burn lasted over six minutes, and the results were excellent.[8] This information would be valuable for Apollo 13, when the LM would be used as a lifeboat and its engine would fire in the same docked configuration to get the crew home quickly.

Seventy hours into the mission, McDivitt and Schweickart re-entered the LM to prepare for Schweickart's EVA, which was shortened to less than an hour due to his in-flight nausea. Though abbreviated, the spacewalk allowed the crew to test the most critical items, including the donning of the spacesuit and backpack, cabin depressurization of both the CM and LM, and opening the hatches of both spacecraft. Schweickart also tested the handrails in case astronauts had to transfer from one spacecraft to the other through the hatches instead of the docking tunnel.

Scott stood up through the open Command Module hatch and retrieved a thermal sample from the Service Module while Schweickart egressed from the Lunar Module to retrieve similar samples from the LM. They were studied after the mission to determine the effects of spaceflight on pieces of glass and other coatings.[9] The EVA was a success.

On the fifth day, McDivitt and Schweickart entered the LM and prepared for a significant mission event: the undocking, separation, rendezvous, and docking of the Lunar Module *Spider* with Command

---

[7] NASA, "Apollo 9 Mission Report," 8-25, 8-26.
Jim McDivitt, interview by author, August 4, 2019.
[8] NASA, "Apollo 9 Mission Report," 1-1, 7-3.
[9] NASA, "Apollo 9 Mission Report," 8-3.

Module *Gumdrop*. The two spacecraft undocked with Scott in the CM using his thrusters to back away at five feet per second. The astronauts in both spacecraft inspected and photographed each other's vehicles while they were within 50 feet. Then a series of maneuvers separated *Spider* and *Gumdrop,* taking the LM 10 nautical miles below and 82 nm behind Dave Scott.[10] McDivitt and Schweickart were the first humans to fly a spacecraft without a heat shield. If they could not rendezvous with Scott, they had no way of surviving reentry in the LM. McDivitt joked that while both spacecraft could land, only the CM could do so safely.[11]

The LM contained a descent stage and ascent stage. After astronauts completed their EVAs on lunar landing missions, the descent stage would remain on the Moon while the ascent stage carried astronauts to rendezvous with the CM orbiting overhead. Four hours into their excursion, McDivitt and Schweickart jettisoned the descent stage to test the flight dynamics and rendezvous capabilities of the ascent stage.

McDivitt told CAPCOM Stu Roosa the ascent engine "is sort of a kick in the fanny compared to the descent engine." As they returned to the CM, McDivitt thought of all the work they put in with FDO David Reed preparing for this moment. The Commander asked the former smokejumper Stu Roosa, "Hey Smokey, is Dave Reed smiling?"

Roosa said, "Well yeah, he's pretty happy, but he's not going to relax until you've finished burning."

"Better not."

The LM primary and backup navigation systems worked well, and *Spider* found its way back to *Gumdrop*, parking about 100 feet away. The astronauts took pictures of each other's spacecraft before successfully docking. The six-hour voyage certified the Lunar Module was ready to go to the Moon.

McDivitt, Scott, and Schweickart spent the last five days of the mission performing photography experiments, three more SPS engine burns, and landmark tracking exercises. Apollo 9 splashed down on March 13 and was a complete success. Now it was time for the next dry

---

[10] NASA, "Apollo 9 Mission Report," 3-2.
[11] Robert Godwin, ed, *Apollo 9: The NASA Mission Reports,* (Ontario: Apogee Books, 1999), 3.

run, which would take the Lunar Module to within 50,000 feet of the lunar surface.

Jim McDivitt, Commander of Apollo 9.
Photo credit: NASA S71-59425.

*Apollo 9 Command Module Pilot Dave Scott stands up through the hatch
of the CM, which is docked with the Lunar Module.
Photo credit: NASA AS09-20-3064.*

First manned flight of the Lunar Module. Photo credit: NASA AS09-21-3213.

12

YOU SURE SEE A LOT OF THE WORLD FROM UP HERE

SNOOPY AND *CHARLIE BROWN*

George Low described the mission sequence building up to achieving President Kennedy's challenge:

> The basic principle in planning these flights was to gain the maximum new experience on each flight without stretching either the equipment or the people beyond their ability to absorb the next step. Too small a step would have involved the risk that is always inherent in manned flight, without any significant gain—without any real progress toward the lunar landing. Too large a step, on the other hand, might have stretched the system beyond the capability and to the point where risks would have become excessive because the new requirements in flight operations were more than people could learn and practice and perfect in available time.[1]

The strategy meant Apollo 10 would be a dress rehearsal, taking the Lunar Module within 50,000 feet of the surface, where powered descent would begin on the actual landing mission. Public Affairs

---

[1] NASA, "What Made Apollo a Success," (Washington, DC: National Aeronautics and Space Administration, 1970), 8–9, https://ntrs.nasa.gov/api/citations/19720005243/downloads/19720005243.pdf.

Officer Terry White referred to the flight as the "Last Step to the Moon."[2]

Apollo 10 was commanded by Tom Stafford, with Gene Cernan as the Lunar Module Pilot and John Young as the Command Module Pilot. All three were veterans of Gemini flights. This crew was the most experienced ever to fly an Apollo mission. It was the third time in space for Stafford and Young and the second for Cernan.

These men had a lot of experience with one another. Tom Stafford and Gene Cernan flew together on Gemini IX where Stafford helped Cernan through a critical situation during Cernan's spacewalk. Additionally, Stafford had been friends with Young for 20 years. "I knew him since we were midshipmen," recalls Stafford. "I was in my first year at the Naval Academy and he was in his first year at Georgia Tech in ROTC [Reserve Officers Training Corps]. We were on the battleship *Missouri* together in the summer of 1949. We got to know each other real well."[3]

Stafford, Young, and Cernan trained together as the backup crew for Apollo 2 before the Fire, then as the backup crew for Apollo 7. On November 13, 1968, one month after the successful flight of Apollo 7, the three veterans were publicly announced as the prime crew for Apollo 10.

Apollo 10 was to simulate the upcoming lunar landing mission as closely as possible, demonstrating rendezvous, docking, and navigation capabilities in lunar orbit.[4] The Sun angle at the landing site would be identical to the conditions during the actual landing.[5]

For the previous flight, the Apollo 9 crew chose the descriptive call signs *Gumdrop* for the Command Module and *Spider* for the Lunar Module. They were personal favorites of Public Affairs Officer Doug Ward.[6] However, some within NASA did not share that opinion, thinking the call signs were undignified. John Young wrote, "We made it worse for them when we announced our names for the Apollo 10

---

[2] NASA, *Roundup*, Vol 8, No 15, May 16, 1969, (Houston: Manned Spacecraft Center, Public Affairs Office, 1969), 1.
[3] Tom Stafford, interview by author, April 7, 2019.
[4] NASA, "Apollo 10 Mission Report," 3-1.
[5] NASA, *Roundup*, May 16, 1969, 1.
[6] Doug Ward, email message to author, March 3, 2020.

modules: *Charlie Brown* for the CM and *Snoopy* for the LM, from the popular *Peanuts* comic strip."[7] Gene Cernan said:

> Everyone on the planet knew the klutzy kid and his adventuresome beagle, and the names were embraced in a public relations bonanza. The intrepid, bubble-helmeted Snoopy, flying his doghouse to the Moon with a red scarf flapping at his neck, became a symbol of excellence, and before the hoopla quieted, that little dog's image was on decals, posters, dolls, kits, sweatshirts, and buttons everywhere. The program had never seen anything like it.[8]

The Peanuts cartoon characters and their creator Charles Schultz enjoyed a long history with NASA. Schultz was an avid supporter of the space program and included space themes in his cartoons. In 1968, NASA approached Schultz asking if Snoopy could be their safety mascot. The "Silver Snoopy Award" was the result, with a pin of the famous beagle presented by astronauts to employees and contractors for exceptional contributions to flight safety and mission success. Schultz drew the design for the silver pins, and each was flown in space before astronauts bestowed this high honor to the deserving winners.

Charles Schultz once told his youngest son Craig that Apollo 10 choosing *Snoopy* and *Charlie Brown* as their call signs was the proudest moment of his career.[9] The Schultz family remained close to NASA over the years. When General Tom Stafford spoke at Space Center Houston's "Thought Leader Series" 50 years after the mission, Schultz's wife Jean was in attendance.

Tom Stafford was a big proponent of including color television on the mission. NASA planned to have a color camera eventually, but Stafford worked with Westinghouse to put a camera together in only a few months,[10] adding a color wheel system to a 12-pound black-and-white camera in time for the mission. The camera was equipped with a

---

[7] Young and Hansen, *Forever Young*, 126.

[8] Eugene Cernan and Don Davis, *The Last Man on the Moon*, (New York: St. Martin's Press, 1999), 197.

[9] Kristen Bobst, "Astronaut Snoopy's 50-Plus Year History with NASA," July 17, 2019, https://www.pbssocal.org/science/snoopy-has-been-working-with-nasa-since-the-late-1950s/.

[10] Tom Stafford, interview by author, April 7, 2019.

zoom lens, enabling long-distance shots of Earth and wide-angle shots within the cabin. The system also included a three-inch black-and-white monitor.[11] The crew used cards with paintings of Charlie Brown in coveralls and Snoopy with his flying ace scarf to calibrate the color camera before broadcasts from space.

SUNDAY MORNING, MAY 18, 1969

Deke Slayton woke the flight crew on launch day at 5:45 a.m. local time. Gene Cernan attended Mass performed by Father Eugene Cargill, a close friend who had flown to the Cape from Houston.[12] After undergoing medical examinations, the crew enjoyed the traditional pre-launch breakfast of steak, eggs, toast, jelly, coffee, and orange juice. George Low was one of several who ate with them. After breakfast, they went to the suit room and donned their space suits while a launch pad team finished loading the propellants. Then they left to board an elevator to the first floor, where a transfer van waited to carry them to the launch pad.[13]

Jamye Flowers, a secretary in the astronaut office, was taking part in a prank on Gene Cernan. The morning of the launch, a NASA team leader named Dave McBride told Cernan that Flowers had something for him to take to the Moon. Cernan agreed, as long as it was small enough to fit in one of his pockets. As they walked down the hallway to the transfer van, there stood Flowers holding a giant stuffed Snoopy. "The gotcha would be on Gene Cernan, that he was going to have to get this very large Snoopy in this very small pocket on the side of his spacesuit," Flowers explained. "The crew were suited up and were coming back down the hallway as I was standing at the door of the crew quarters, just inside with the Snoopy in my arms. Right at the last minute, Dave McBride gave me a push, and I ended up out in the hallway," she said.[14]

---

[11] NASA, *Roundup*, May 16, 1969, 1.
[12] Cernan and Davis, *Last Man on the Moon*, 167.
[13] Details were collected from the pre-launch commentary by KSC PAO Chuck Hollinshead.
[14] Jamye Flowers Coplin, interview by Rebecca Wright for JSC Oral History Project, November 12, 2008, 23.

Stafford was the first to walk by, patting the giant beagle on his nose, a scene captured in a famous photograph. Young did the same. Flowers describes what happened next:

> Out of the corner of my eye I could see Captain Cernan heading in my direction, and he was a man on a mission at that point. So, I knew that I was in trouble. But he came and instead of patting Snoopy on the nose, he turned the prank into a gotcha on me because he grabbed me and Snoopy and tried to get us in the elevator. Tom Stafford still says it is one of his favorite photos, if not his favorite.[15]

After the transfer van delivered the crew to the launch pad, they rode the elevator inside the launch tower to the 320-foot level, then walked across the gantry to the White Room surrounding the Command Module hatch. The gantry provided an excellent view of the beach, bumper-to-bumper cars, American flags, and the packed VIP stands three miles away.[16]

Once inside the White Room, they were met by Guenter Wendt's closeout team that assisted them into the Command Module high atop Pad B of the Launch Complex 39. This was the only time NASA used Pad B for a Saturn V launch. The primary site, Pad A, was being prepared for the Apollo 11 stack which would be rolled out before Apollo 10 reached lunar orbit.

Stafford, Cernan, and Young were launched 49 minutes past noon local time on May 18, 1969. A few minutes after liftoff, the crewmen were slammed into their seat harnesses as the first stage engines shut off before the second stage took over. Stafford said, "We went from nearly five Gs to zero in a fraction of a second, flying toward the control panel. The shutdown felt like being in a train wreck."[17]

Cernan asked CAPCOM Charlie Duke, "Are you sure we didn't lose *Snoopy* on that staging?"

Duke replied, "No, I think *Snoopy* is still there with you. You're looking good."

---

[15] Jamye Flowers Coplin, interview by Rebecca Wright, 23.
[16] Cernan and Davis, *The Last Man on the Moon*, 198.
[17] Thomas P. Stafford, *We Have Capture: Tom Stafford and the Space Race*, with Michael Cassutt, (Washington: Smithsonian Books, 2002), 123.

The veteran crew could not hide their excitement.

STAFFORD: It looks good to be back up here, Charlie.
YOUNG: Just like old times. It's beautiful out there.
DUKE: You guys sound ecstatic.
STAFFORD: Man, this is the greatest, Charlie.
CERNAN: Charlie, babe. It's fantastic, babe, really!

Apollo 10 successfully entered Earth orbit, circling the globe at an altitude of 100 nautical miles. They checked their systems after the jarring ride, and all looked good. Two and a half hours into the mission, it was time for Translunar Injection (TLI) as they passed over Australia at night. TLI was a six-minute burn converting the present circular orbit into a large elliptical orbit, which at its high point would intercept the Moon's path. If the engine did not fire long enough to connect with the lunar gravitational field, orbital dynamics would return the spacecraft to the same place the burn took place, 100 nautical miles above Earth. From there, the crew would perform a maneuver to bring them down from orbit.

The astronauts were flying upside down in the darkness and could see the lights of Sydney as TLI began. Cernan describes what happened next:

Then the mighty rocket fired, and Sydney vanished. The Aussies were stunned by the brilliant new star that lit up their night sky. Unlike Gemini, where dawn arrived at a relatively slowpoke pace, we now crashed into daylight, climbing away from Earth at incredible speed. "What a way to watch the sunrise, I marveled."[18]

The TLI began as expected, but a few minutes into the burn, the crew experienced extreme vibrations. The Saturn's third stage shuddering was so violent all three astronauts expected the mission would be aborted. "We all figured the flight was over right there," said John Young. "From three minutes on, we held our breath."[19]

---

[18] Cernan and Davis, *The Last Man on the Moon,* 203.
[19] NASA, "Apollo 10 Technical Crew Debriefing," 5-2.

Stafford had his hand on the abort lever the whole time but did not twist it. "We rode that wildly bucking creature through the entire burn, and Tom never removed his gloved hand from the abort handle," wrote Cernan.[20] After six minutes, the S-IVB third stage shut down, the vibrations stopped, and the crew was off to the Moon.

John Young changed seats with the Commander after TLI, taking control of the spacecraft. As the Command Module Pilot, Young separated the CSM from the third stage which still carried *Snoopy*. He maneuvered away, pitched the CSM 180 degrees, and slowly went back to dock with the Lunar Module. Cernan pulled out the television camera and filmed the first color TV broadcast from space. After a smooth docking, Young backed away from the third stage, pulling the LM with him.

STAFFORD: *Snoopy* is coming out of the doghouse.
DUKE: Roger. And we got the TV.

Live color television showed spectacular scenes of the Saturn's third stage drifting away from the camera.

STAFFORD: I wish you'd tell Dr. von Braun, Lee James, Kurt Debus, and Rocco Petrone thanks a lot for all the people who worked on the great ride.
DUKE: Roger. We're going out to the networks now. They probably heard it. We'll pass it on, though.
STAFFORD: A few thousand people worked on that machine, and we sure appreciate it.
DUKE: Roger. It looked beautiful from here.

Gene Cernan said, "Charlie, we're looking for the Earth right now. We'd like to show it too, but we can't find it."
Duke said, "Roger. It's down there somewhere."
The S-IVB was slowly drifting away from the docked CSM and LM but was still too close for comfort. The Service Module SPS engine performed a three-second braking burn to let the S-IVB move ahead. Once forward of the spacecraft, the third stage discharged its remaining

---

[20] Cernan and Davis, *The Last Man on the Moon,* 203.

propellants through the engine in a propulsive vent, adding more separation.

Then it was time for another TV broadcast. They were five hours into the mission and 21,000 nautical miles from home.

> STAFFORD: I'm looking right at the good ol' US of A there.
> DUKE: Hey, it's really beautiful, Tom. It's coming in great.
> STAFFORD: You ought to see it up here, Charlie.
> YOUNG: We got the whole globe there.

Stafford zoomed in on the Rocky Mountains and California. Duke was blown away by the color images. He enthusiastically commented, "This has got to be the greatest sight ever."

Stafford agreed, "You ought to see it up here."

Cernan paused as he searched for words. "Charlie, it's so hard to describe. You can go right up past Alaska, and you can see the polar caps. It's incredible."

Duke mentioned the beautiful colors. Cernan added, "And the blackest black that you ever could conceive is the setting for all this."

> DUKE: You guys are really giving us a great show. This is
> fantastic.
> STAFFORD: We just want to thank all the people who helped
> get us up here, Charlie. That includes the taxpayers too.
> CERNAN: You blink your eyes and you look out there at that
> ball and you know it's three-dimensional, but it is just sitting
> out there in the middle of nowhere and it's unbelievable!
> DUKE: We are getting a real idea now, for the first time, of
> what you are really seeing up there. The colors on the oceans
> are just as blue as they can be, and real white clouds all over,
> and the land is a real deep brown, almost a reddish brown.

The crew continued describing the geography and cloud patterns they observed. A few minutes later, Cernan said, "We are looking right at you. Looks like you may have a few clouds there right now in Houston." Duke, sitting in a smoke-filled Mission Control, added, "It's sort of a constant overcast here in the MOCR."

Cernan summarized his view of the planet. "Just for the record, it looks to me like a pretty nice place to live."

MAGELLAN

John Young had a lot of fun calling Mike Collins "Magellan" on their Gemini X flight due to all the navigation duties Collins performed. Now it was Young's turn to test his skills. He had to periodically realign the spacecraft's guidance system the same way sailors have navigated for centuries, using a sextant and the stars. Young's tools included a 28-power sextant and a one-power wide-angle telescope. Neil Armstrong once said, "NASA is probably the only organization in history that's been sold a one-power telescope."[21]

The scanning telescope had a 60-degree field of view, allowing astronauts to see entire constellations and recognize the 37 stars designated as navigation stars.[22] It was similar to finder scopes used by backyard astronomers. NASA selected navigation stars located away from the Sun; the Command Module Pilot (CMP) could easily detect them when performing alignment checks.

Each star was numbered and pre-programmed into the guidance computer. The CMP entered a star's information. Once he found it in the wide-field telescope, Young slid over to the 28-power sextant, which he considered "ingenious."[23]

The sextant measured the angle between a known landmark and a star. It contained mirrors and two lines of sight. One line of sight was stationary and was placed on the desired Earth or lunar landmark, such as the horizon. To do this, Young had to maneuver the whole spacecraft using the attitude controller until the chosen landmark was in the center of the eyepiece.

The second line of sight was adjustable, independent of the landmark line of sight, and could be linked to the scanning telescope.[24] It was maneuvered by an optics hand controller. Young simultaneously used the optics controller to keep the star centered in the eyepiece while using the attitude controller to maneuver the spacecraft until the

[21] Neil A. Armstrong, interview by Stephen E. Ambrose and Dr. Douglas Brinkley for NASA Johnson Space Center Oral History Project, September 19, 2001, 63.
[22] David Hoag, "Apollo Navigation, Control, and Guidance Systems: A Progress Report," April 1969, accessed January 8, 2020, https://web.mit.edu/digitalapollo/Documents/Chapter6/hoagprogreport.pdf, 13.
[23] Young and Hansen, *Forever Young*, 128.
[24] Young and Hansen, *Forever Young*, 128.

landmark came into view. When the star and landmark were superposed, he pressed the "Mark" button, telling the guidance computer to measure the sextant angle.[25]

The information updated the computer, which was connected to the Inertial Measurement Unit (IMU), the heart of the Apollo Guidance System. The IMU was contained in a sphere the size of a beach ball. At its center was a "Stable Member" platform held in place by three gyroscopes. The platform also had three accelerometers to measure any change in velocity in the yaw, pitch, and roll axes.

The stable member was mounted inside the IMU by three gimbals, each aligned 90 degrees from the others. This allowed the platform to remain in the same orientation in reference to the stars while the spacecraft rotated around it. As precise as the IMU was, it gradually drifted over time and had to be corrected through periodic star sightings.

Refining the platform alignment was similar to balancing a checkbook in the days before online banking. Using the previous bank statement as a correct starting point, one would add the deposits and subtract the checks throughout the month, giving a good idea of how much money was in the account. When the following bank statement arrived with exact debits, credits, and service charges, their figure would be adopted as the new correct balance unless there was a large discrepancy.

On the Gemini X flight, Mike Collins had difficulty recognizing Earth's horizon; it was fuzzy due to the light scatter in the atmosphere, but precision was required for an accurate sighting. Young and other Apollo navigators experienced the same phenomenon. The MIT Instrumentation Laboratory developed a simulator to replicate the fuzzy horizon. Young recalled, "Before each mission, including Apollo 10, the navigator-astronaut came to the Instrumentation Laboratory to train on this simulator. With practice, I got to the point where I could duplicate my sighting point within a few kilometers over the desired range of distances to Earth."[26]

---

[25] Charles Duke and Michael Jones, "Human Performance During a Simulated Mid-Course Navigation Sighting," (Master's thesis, Massachusetts Institute of Technology, 1964,) accessed January 8, 2020, https://authors.library.caltech.edu/5456/1/hrst.mit.edu/hrs/apollo/public/archive/1130.pdf, 1–4.
[26] Young and Hansen, *Forever Young*, 129.

The practice came in handy early in the mission when Young used Earth's horizon as a landmark. He said, "All in all, the navigation part of my assignment was at times highly challenging but always tremendously interesting."[27]

The crew finished their dinner ten hours into the flight. They were 50,000 nautical miles from Earth, traveling at 8,500 feet per second and gradually slowing. Next on Young's list was putting the spacecraft in a slow roll to balance the temperature. This Passive Thermal Control (PTC) was affectionately known as the "Barbeque Roll." The spacecraft slowly rotated around its long axis like a rotisserie chicken. Otherwise, the side facing the Sun would overheat while the side away from the Sun would quickly freeze.

WEATHER AND NEWS

When the astronauts woke for day two, they were 92,000 nautical miles away from home, and their velocity had dropped to 5,800 feet per second due to Earth's gravity. John Young gave CAPCOM Charlie Duke a detailed weather report from high above the planet.

Duke replied, "Thank you, Apollo 10. The only thing missing was the music."

Young said, "That's a special effect we're not carrying today."

After reading flight plan updates, Duke was joined by Bruce McCandless to tag team the morning news.

> **DUKE:** We're ready with a summary of news and sports as compiled by your friendly third floor astonisher Jack Riley and his office. Are you ready?
> **CERNAN:** Man, we're just about ready for anything.
> **DUKE:** Roger. This is a news team of McCandless and Duke. Newspapers, television, and radio are concentrating on the flight of Apollo 10. The *Houston Post* banner read "Apollo 10 Out of This World," and for the first time in memory, the entire front page of the *Post* is all space news. The newswires are commenting on the quality and quantity of the TV transmissions yesterday.

---

[27] Young and Hansen, *Forever Young*, 129.

McCANDLESS: Senator Barry Goldwater paid surprise visits to the Stafford and Young homes yesterday. He said he came to Houston because he had been to the Cape before, and each time the launch had been postponed. Other than the Apollo 10 mission, the world has been relatively quiet.

Duke and McCandless continued their snappy, well-timed routine for over three minutes, discussing items such as the Preakness horse race, each crewman's horoscope, and the weather.

DUKE: And this finishes the first annual McCandless/Duke radiocast. Over.
McCANDLESS: Roger. Good morning, Charlie.
DUKE: Good morning, Bruce.

The performance drew laughter from flight controllers and rave reviews from the Moon-bound audience. Stafford said, "You guys are too much down there. That's fantastic."

Young added, "Boy, you outdo me. I quit. You can give the weather next time, too."

Cernan said, "You're going to put someone out of business down there if you don't watch out."

Duke quickly replied, "Maybe you guys."

One of the next items on the schedule was a waste-water dump. Knowing telescopes around the world were tracking the spacecraft, they announced the time of the dump 100,000 miles away from Earth so it could be observed.

About an hour later, the crew had a message for Duke and McCandless.

CERNAN: You guys were so good to us with the news this morning that we thought we'd bring you a little disc jockey work from up here if you're prepared.
DUKE: Roger.
CERNAN: This is Tom and John on the guitar and the three of us singing.

The crew held their tape recorder to the mic and played "Up, Up, and Away" by the 5ᵗʰ Dimension. After the popular song ended, Duke said, "Hey, that was really beautiful. Somebody's voice is changing, though, or you stowed somebody away up there." He continued, "It was really great, you guys. Y'all have been practicing a lot."

Young replied, "We had trouble stowing the bass drum aboard but other than that, it came out pretty well."

Tom Stafford said, "The five psi makes your voice a little higher, Charlie." His comment caused widespread laughter inside Mission Control. The oxygen pressure inside *Charlie Brown* (five psi) was one-third the pressure of air we breathe on Earth.

> **PUBLIC AFFAIRS OFFICER** *(JACK RILEY): That's another space first for Tom Stafford, although it probably won't rank as high in the technical annuals as the first space rendezvous he and Wally Schirra performed.*

At about 3:00 p.m. Houston time, the crew began another broadcast from space. Bruce McCandless was CAPCOM. After giving a weather report, they zoomed in on Earth. Cernan said, "It's a beautiful sight. We're sitting here, and it's almost like science fiction looking back at it, Bruce."

Young chimed in. "I'm voting for the world being round, if there are any dissenters."

McCandless said, "Roger. We'll record your vote on that issue."

Then Stafford pointed the camera inside the spacecraft while John Young displayed a series of cards. The first was the busy Apollo 10 mission logo. He joked, "We were going to put some more things in it, but we just ran out of time." Mission Control broke out laughing.

Young was handed a card picturing Charlie Brown. "This is our other emblem," he said.

Cernan asked McCandless, "Do you see any resemblance between the card and the guy holding the card?"

CAPCOM replied, "Now that you mention it...."

They gave Young a card displaying Snoopy. Cernan said, "Boy, he's been quiet for two days; he's going to get a chance to do a little woofing here in the next couple of days."

Cernan took the camera, showing Young upside down with Stafford right side up. The pair were usually reserved on Earth but were hamming it up in space. Young said, "One of our problems is trying to figure out which way is up and which way is down."

Stafford added, "You have your choice. If you don't like things right side up, you can go upside down."

McCandless cracked up along with Charlie Duke, Deke Slayton, and Harrison Schmitt standing behind him. McCandless replied, "We've got one of you in each direction."

Tom Stafford expressed what everyone could see in living color, "It's really a ball up here living in zero-G, believe me."

Young said, "It's the only way to fly."

The Commander said, "Once you get going, the cost for individual passenger mile becomes rather reasonable."

Then Stafford began moving Young up and down by pushing the top of his head. From behind the camera, Cernan commented, "You might notice the dynamics here."

Young said, "I just do whatever he says."

McCandless said, "Tom, the flight surgeon here wants you to be sure you log all your exercises."

The following afternoon, the voyagers again described the cloud patterns from their lofty perch. Duke said, "You guys are giving us a great weather report."

Young said, "You sure see a lot of the world from up here.

"Yeah, like maybe all of it."

Tom Stafford (left) maneuvering an upside-down John Young.
Screen capture from NASA television.

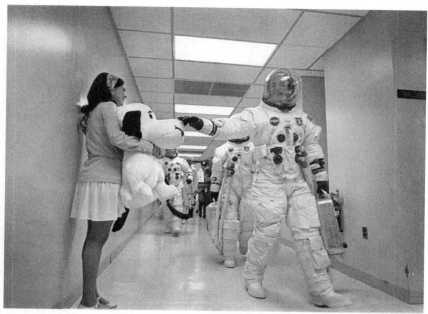

*Jamye Flowers holds Snoopy as Tom Stafford greets him on the way to the
transfer van and launch pad. Gene Cernan is visible behind Snoopy.
Photo credit: NASA 69-H-801, Kipp Teague, scan by Ed Hengeveld.*

*Gene Cernan, Tom Stafford, and John Young enjoying the company of
Vice President Spiro Agnew (second from left) and Snoopy.
Photo credit: NASA KSC-69P-348, Kipp Teague, scan by J.L Pickering.*

Snoopy and Charlie Brown accompany CAPCOM Charlie Duke as Apollo 10 prepares for TLI two and a half hours into the mission. Photo credit: NASA.

13

HEY, SNOOP, AIR FORCE GUYS DON'T TALK THAT WAY

TO THE MOON

Charlie Brown, *Snoopy*, and their three humans continued to the Moon, arriving on the mission's fourth day. Young performed a retrograde Lunar Orbit Insertion burn lasting six minutes to slow them enough to be captured by lunar gravity. The result was a 60 by 170 nautical mile elliptical orbit. A few minutes before the burn, the crew got their first breathtaking glimpses of the Moon. Commander Stafford told them to keep their heads inside the cockpit.

Once the burn was completed, they could no longer contain themselves. "We became like three monkeys in a cage, scrambling to the windows to get a close look at this big gray thing turning below us," wrote Cernan. "We saw scarred mountains and valleys, deep craters and deeper canyons, rilles and gullies, possible ancient volcanoes that were white on the outside and black inside, and circular craters of all sizes, from baseball fields to Rhode Island. But not a single sign of life."[1]

The LOI burn occurred on the far side of the Moon, where there was no communication with Houston. Mission Control waited quietly to see if the burn had gone according to plan.

---

[1] Cernan and Davis, *The Last Man on the Moon*, 210.

They regained radio contact right on time. Stafford told CAPCOM Charlie Duke, "You can tell the world that we have arrived." The crew showered the Flight Dynamics Officers with praise as their calculations delivered the astronauts within one mile of the planned orbit. As they described the lunar features, Gene Cernan said, "Charlie, it might sound corny, but the view is really out of this world."

Duke laughed and said, "We had a couple of comments from the back row that I won't repeat."

The future landing site on the Sea of Tranquility was still in darkness as they flew over for the first time, as it would be at this point of the actual landing mission. However, the Sun had risen over some prominent landmarks leading up to it. One was Rima Hypatia, a long rille parallel to the flight path. (Rille is German for "groove.") The astronauts named it "US Highway 1" after the well-known road running up and down America's east coast. The Commander wasted no time in naming something after his native state, Oklahoma.

> STAFFORD: The feature we called "US 1" stands out real well.
> It disappears in the darkness right by Moltke. And in the area
> over to the right there's no doubt there's been some
> volcanism in there, and that's what we term the "Oklahoma
> Hills."
> YOUNG: I knew he'd name something "Oklahoma Hills."
> DUKE: You notice he got that out on the first revolution, too.

The crew was amazed at the sights caused by a phenomenon called earthshine. Earth reflects sunlight onto the dark portions of the Moon, creating an eerie scene. It is similar to the Moon reflecting light on Earth at night, which creates shadows and forces militaries to schedule covert operations when the Moon is not in the sky. Cernan said, "It's amazing how well you can, once you find the landmark, navigate in earthshine across the surface of the Moon. It seems to be very well-lit from our altitude here…. The minute we go out of sunlight in the darkness ourselves, the Moon then glows right at us."

Stafford said, "In earthshine, you can see way down in the craters. You can see the shadows in the craters from the earthshine, but the more you become adapted to it, it's phenomenal the amount of details you can see."

Two revolutions later, Young performed a 14-second maneuver that circularized the orbit at 60 nautical miles. Next came the first color television broadcast from lunar orbit. It was an impressive half-hour tour of the path to the landing site, which was now barely illuminated by the Sun.

The next day would be "Rendezvous Day," simulating the landing day procedures for Apollo 11. After the broadcast, Stafford and Cernan prepared to float through the docking tunnel to check out *Snoopy*. As they opened the tunnel hatch, the men were greeted by a stream of insulation from the hatch cover. After cleaning up the mess, Stafford and Cernan checked the multiple Lunar Module communication modes and antenna options, making sure they were ready.

Meanwhile, Young remained in *Charlie Brown* performing lunar landmark tracking exercises. Flight controllers needed to understand lunar orbits as much as possible. This task was complicated by varying mass concentrations hidden under the lunar surface. The controllers selected small craters along the flight path and predicted the exact times the spacecraft would fly over them on future revolutions around the Moon. "The concept was to make sure we had a good relative understanding of the orbit vs. the landing site," explained Guidance Officer Steve Bales.[2] Young marked the time he approached each landmark at specific angles, and the FDOs compared the data with the predictions. The conclusions helped refine their knowledge of a spacecraft's orbit.

The ability to precisely predict future orbits would be necessary for the next mission, which would attempt to land following the same ground track as Apollo 10. At the beginning of its descent, the Apollo 11 LM guidance computer would have to know where it was, what direction it was going, and how fast it was traveling. It would be too late for Mission Control to send that information after the spacecraft appeared from behind the Moon only a few minutes earlier. Those details had to be predicted hours in advance and sent up on the previous revolution. Hence, the importance of landmark tracking exercises.

After Stafford and Cernan completed their tests in *Snoopy* and returned to *Charlie Brown*, CAPCOM Joe Engle asked them about the

---

[2] Steve Bales, email message to author, March 30, 2020.

insulation. Cernan replied, "Would you believe we've been living in what you might call snow for three days? And we found out where the rest of it is. It's in our good friend *Snoopy*."

Engle asked if inhaling it was causing any problems. Cernan said, "I didn't have to worry about inhaling it. I *ate* my way through."

Stafford said, "Your throat feels lousy and your nose wheezes a little bit."

Cernan said, "That should be a space first—snow on the Moon."

RENDEZVOUS DAY

Day 5 was prime time for the Apollo 10 mission as the crew would simulate the operations for landing day on Apollo 11. After breakfast, Cernan and Stafford went inside *Snoopy*, meticulously cleaned the insulation, powered up the Lunar Module, checked out systems, and prepared for the busy day.

Communications problems quickly reared their ugly head. Then they had problems venting the tunnel because of a valve issue. Additionally, the docking interface caused slippage of three degrees between the two vehicles. These issues forced a delay in undocking.

Finally, at 2:00 p.m. on the far side of the Moon, John Young released the twelve docking latches, and the two vehicles drifted apart. They flew in close formation for half an hour until Young performed a separation maneuver as they neared the Apollo 11 landing site. As a result, *Charlie Brown* moved lower and ahead of *Snoopy*, clearing the way for *Snoopy's* upcoming Descent Orbit Insertion (DOI) maneuver half of a revolution later. The DOI occurred on the back side of the Moon, opposite the landing site. The one-minute DOI burn slowed the LM, making its orbit elliptical with the low point only 50,000 feet above the surface where the Apollo 11 descent would begin.

After the DOI maneuver, Young watched *Snoopy* through his optical system as the LM dropped lower and lower. Cernan was glad Young was in *Charlie Brown*:

> As we put more distance between the two spacecraft, I was
> comforted knowing that John Young was up there in our
> Command Module. Those CM guys were some of the best
> fliers in the business. We had worked and trained together so

closely that John always knew exactly what we would be doing and precisely where we were. We wouldn't have wanted anyone else as our lifeguard. Ask any astronaut who walked on the Moon about the guy who stayed in orbit, and you will hear nothing but high praise.[3]

Moments before the acquisition of signal with Houston, Stafford and Cernan witnessed an incredible earthrise. Cernan exclaimed, "Look at that! Look at the Earth! Look at the Earth!" They continued to dip toward the low point of their orbit.

CERNAN: We is down among 'em, Charlie.
DUKE: Roger. I hear you're weaving your way up the freeway. Can you give me a post-burn report?
CERNAN: Yeah, as soon as I get my breath!

They tested the landing radar. It achieved lock-on with the lunar surface well above 50,000 feet, sooner than expected. They also photographed the flight path toward the landing site and described it in detail.

Stafford reported, "Charlie, it looks like we're getting so close all you have to do is put your tail hook down and we're there."

Duke said, "Hey, *Snoop*, Air Force guys don't talk that way."[4]

Young was watching them from high above: "OK, Houston. I've got them in the optics now. They're fantastic. Boy! Are they down there among them!"

"Rog. I bet it looks like they're really hauling the mail."

Cernan said, "Surprisingly enough, Charlie, it really doesn't look like we're moving too fast down here. It's a very nice pleasant pace."

Snoopy hit the low point of his orbit over Mount Marilyn, named after Jim Lovell's wife, where the Apollo 11 descent would begin. Soon they passed by Weatherford Crater, named after Stafford's hometown in Oklahoma. Then came the prominent landmarks Maskelyne, Duke

---

[3] Cernan and Davis, *The Last Man on the Moon*, 214.
[4] Stafford and Duke graduated from the US Naval Academy, but both were Air Force officers. Naval fighter pilots like John Young and Gene Cernan land on aircraft carriers in planes with tail hooks that engage steel arresting cables.

Island, Boot Hill, Sidewinder Rille, Diamondback Rille, Moltke, US 1, followed by the landing site.

Some of the key markers were rilles (trenches). Cernan described them from close range. "Diamondback Rille is very easy to see. These rilles look like they may be as much as a couple of hundred feet deep and very smooth. The surface actually looks very smooth, like very wet clay, but smooth with the exception of the bigger craters."

He grabbed a camera as Stafford continued the commentary. "Sidewinder Rille is round along the edges. It is flat under, and smooth in the bottom. The ridges are definitely rounded and it doesn't look like the sides are upturned."

Cernan followed, "The best description I can give you of these rilles is of a dry riverbed out in New Mexico or Arizona."

If left alone, *Snoopy* would start climbing again, returning to the location of its DOI burn on the opposite side of the Moon. The LM would keep moving farther ahead of *Charlie Brown,* which remained in its circular orbit 60 nautical miles high. One of the mission goals was to replicate the ascent from the lunar surface after the landing. The ascent stage of the LM would approach the Command Module from below and behind. For *Snoopy* to simulate the approach and rendezvous, it had to change its flight path, letting *Charlie Brown* move ahead of it.

Therefore a "Phasing Maneuver" was performed, burning the LM descent engine 40 seconds which increased its altitude on the other side of the orbit to 190 nautical miles. The higher trajectory allowed *Charlie Brown* to pass underneath *Snoopy* as it soared overhead. The next revolution brought *Snoopy* down near the Moon again in the correct position to chase *Charlie Brown* from below and behind.

The Lunar Module consisted of two stages bolted together. The descent stage carried most of the weight and had a throttleable engine that would lower it to the Moon on landing missions. After the exploration of the lunar surface was complete, the descent stage would act like a launch pad while the ascent stage, carrying the crew, would return to orbit and rendezvous with the Command Module. The previous mission, Apollo 9, took the LM away from the CM while in Earth orbit, then the ascent stage separated from the descent stage and rendezvoused with the Command Module. The same maneuver was planned for *Snoopy* in lunar orbit.

On the far side of the Moon, *Snoopy* began its dive toward the second low-level pass over the landing site. At 6:00 p.m., an eerie noise grabbed the crew's attention.

Cernan asked Stafford, "That music even sounds outer-spacey, doesn't it? You hear that? That whistling sound?"

Young also heard it and radioed *Snoopy*. "Did you hear that whistling sound, too?"

Cernan said, "Yeah. Sounds like outer-space-type music. Isn't that eerie, John?"

"Yes, I was going to see who was outside."

As Stafford and Cernan approached the landing site for the second low-level pass, they followed the checklist and flipped switches preparing to separate from the LM descent stage. An Abort Guidance System switch was accidentally changed twice and ended up in the wrong position, telling *Snoopy* to start looking for *Charlie Brown* immediately.[5] The LM abruptly went out of control.

"We were suddenly bouncing, diving, and spinning all over the place as we blazed along at 3,000 miles per hour, less than 47,000 feet above the rocks and craters," wrote Cernan. "The spacecraft radar that was supposed to be locking on to *Charlie Brown* had found a much larger target, the Moon, and was trying to fly in that direction instead of the orbiting Command Module."[6] John Young and Mission Control could only listen as Stafford and Cernan fought to stop the gyrations.

Stafford took manual control of *Snoopy*, jettisoned the descent stage, and soon had the ascent stage in check. Next, they fired the engine to simulate an ascent from the lunar surface and chased down *Charlie Brown*.

Snoopy eased beside *Charlie Brown*, and they docked soon after 10:00 p.m. Stafford reported to CAPCOM Joe Engle, "Hello, Houston. *Snoopy* and *Charlie Brown* are hugging each other."

Two hours later, all three men were back in the Command Module. They stored everything they did not need in *Snoopy*, closed the hatch, and undocked. Oxygen remained in the tunnel at a pressure of five

---

[5] Stafford and Cassutt, *We Have Capture*, 130.
[6] Cernan and Davis, *The Last Man on the Moon*, 217.

pounds per square inch. The pressure sent the LM away quickly in the direction of the Sun, making it impossible for the crew to see it.

Stafford said, "Cabin pressure's holding. *Snoop* went someplace."

Young said, "Man, when he leaves, he leaves!"

Stafford continued, "Joe, he took off so fast, he's gone! He went right into the Sun. We don't have any idea where he went. He just went boom and disappeared right into the Sun."

Stafford wanted to make sure they did not collide with *Snoopy*. He began talking through the situation with Houston:

> STAFFORD: Let's take a quick look at these orbital mechanics. When we separated, he had that five psi in the tunnel and he took off like a scalded rock straight up. OK?
>
> ENGLE: Roger that.
>
> STAFFORD: Now he's up above us someplace, and I don't know where. Now, do you want us to thrust down?
>
> ENGLE: Stand by, *Charlie Brown*. We're running this thing through right now.
>
> STAFFORD: Yeah. Because we don't want to see *Snoopy* come back here with a full head of steam.

Engle called Stafford with an answer, but misidentified himself. "*Charlie Brown*, this is *Snoopy*. Or, *Charlie Brown*, this is Houston."

The quick-witted Commander replied, "Yeah, I hope this is Houston. We're going to try to pick *Snoop* up on our VHF ranging. But go ahead."

Engle laughed and asked, "You didn't leave anybody in there did you, Tom?"

"No. I don't think so."

Following Houston's advice, Stafford briefly thrust downward two feet per second to give them more separation from the Lunar Module. When *Snoopy* was 2,000 feet away from *Charlie Brown*, Houston remotely fired the LM ascent stage engine until its propellants were depleted four minutes later.

Cernan asked, "Is he really going to the Sun?"

Engle said, "Well, he's going in that general direction."

"I feel sort of bad about that because he's a pretty nice guy. He treated us pretty well today."

Before the astronauts went to bed, their boss Deke Slayton got on the communications loop to congratulate them on a job well done:

> SLAYTON: That was a beautiful job today. If you do half that well tomorrow, we'll let you come home.
> YOUNG: We'll do better than that tomorrow.
> CERNAN: There's a lot of people who did a good job. And I'll tell you, these vehicles, that little *Snoopy* was a real winner.
> YOUNG: And big *Charlie Brown* is no slouch either.

DAY 6

The next morning, CAPCOM Jack Lousma updated the crew on their friend *Snoopy*. "You might be interested to know that the LM ascent stage is 23,000 miles from the Moon, heading straight up at 5,400 feet per second." Lousma continued, "I've got a congratulatory message here. It says, 'Congratulations on doing what I've been trying to do for a long time. Signed, Red Baron.'"

Cernan photographed targets of opportunity designated by geologists, while Stafford and Young spent the busy day working closely together performing landmark tracking exercises. A NASA Technical Note on lunar landmark locations describes the process:

> In a typical tracking sequence, a set of five sightings [called marks] is taken as the spacecraft passes over the landmark. The first mark is taken when the approaching spacecraft is approximately 35 degrees above the landmark local horizon; the third mark is taken when the landmark is at the spacecraft nadir [directly below the spacecraft] and the fifth mark is taken when the receding spacecraft is again at 35 degrees. The second and fourth marks are spaced evenly between these three marks. The optimum time interval between marks is approximately 20 to 30 seconds for the nominal 60-nautical-mile-high circular orbit.[7]

---

[7] Gary Ransford, Wilbur Wallenhaupt, and Robert Bizzell, "Lunar Landmark Locations—Apollo 8, 10, 11, and 12 Missions," (Houston: Manned Spacecraft Center, 1970,) https://ntrs.nasa.gov/api/citations/19710002567/downloads/19710002567.pdf, 1.

The ground predicted the time of the spacecraft's pass over each landmark at the designated angles and compared those predictions with actual results. The process took a coordinated effort between Stafford and Young, with the Commander flying the spacecraft and looking out one window, alerting Young of the upcoming targets coming into view of his sextant.

One of the most significant targets was Landmark 130, a crater 750 meters in diameter located 20 kilometers north-northwest of the Apollo 11 landing site. 130 sits on the southwestern edge of Sabine D, a three-kilometer-diameter crater later named "Collins" by the International Astronomical Union (IAU) in honor of astronaut Mike Collins. For a fine-tuned target, John Young used a rockslide on the northeast wall of 130. Collins also used 130 as a landmark during Apollo 11, zeroing in on a small crater on its northern rim.[8] Stafford and Young drew high praise from Mission Control over their fine work during the tracking exercises.

In between landmark sightings, the crew noticed a bright object with a different appearance than a star. They realized it could be the descent stage of *Snoopy*, which remained in its elliptical orbit after being jettisoned during the previous day's heart-stopping moments. While the ascent stage was embarking on a permanent orbit around the Sun, the descent stage crossed the CM's circular orbit. Cernan viewed it through a monocular and confirmed it was the bottom half of *Snoopy*. "Hey, that's it, babe. You can see the legs. You can see it tumbling."

The tracking exercises and photography lasted until 9:00 p.m. The Trans-Earth Injection was scheduled for 5:30 the following morning, so the crew took a brief rest period of three hours.

At midnight CAPCOM Joe Engle put in a wake-up call to the crew. He then asked for a crew status report, which typically included details of what food each astronaut had eaten the previous day. Cernan replied, joking, "This is Gene Cernan calling from the Moon. As I look around, there are three of us: John Young, Tom Stafford, and myself. And their status has been fairly confident. Can we help you?"

"You got me," said Engle.

---

[8] Ransford, Wallenhaupt, and Bizzell, "Lunar Landmark Locations," 8.

The crew had a midnight television transmission as they passed over the landing site. During the broadcast, *Snoopy's* descent stage made a guest appearance. They tried to show it on television, but the camera could not distinguish it from the lunar background. Stafford estimated it was no more than 10 miles from them. "Seeing what I saw yesterday, we sure don't like to be around here playing footsy with that rascal."

Deke Slayton jumped into the conversation. "You guys treated him so badly on staging, he's out to get you."

> **ENGLE:** We're trying to figure out what *Snoop's* doing right now.
> **STAFFORD:** I know it's highly improbable, a collision, but it would sure ruin your whole day if it ever happened.
> **ENGLE:** Ol' *Snoop's* just a devoted old hound dog, Tom. He'll probably try and follow you back home.
> **STAFFORD:** Just as long as that rascal doesn't sniff too close.

The flight controllers concluded *Charlie Brown* was lapping the descent stage, and this encounter was as close as they would come before it was time to leave for home.

Snoopy was not their only concern. A pump associated with the first fuel cell stopped working the previous day. The spacecraft was built with redundancy, backup systems, so one failure would not cause a problem. In this case, there were two other fuel cells for the return to Earth. Early in the morning, on the far side of the Moon, a warning light for the second fuel cell indicated a fluctuating temperature. They reported it to Mission Control when they reacquired communications, but by then it had stabilized. The fluctuations returned when the spacecraft went into darkness, but the fuel cell was functioning normally.

They spent the last two hours of lunar orbit configuring their spacecraft for the Trans-Earth Injection. The two-minute-and-forty-four-second burn would occur on the far side of the Moon. Young maneuvered the spacecraft to the correct attitude, then checked its alignment with the stars. The 20,500-pound thrust Service Propulsion System engine fired right on time, and they headed home.

The last full day of the mission began with the crew playing "Come Fly with Me" on the communications loop. Gene Cernan was the disk jockey:

> Good morning, good morning! This is Tom, John, and Gene
> broadcasting again from approximately 140,000 miles out in
> the universe. It's a beautiful day out here, and it appears that
> it might be a beautiful day down in mother Earth country.
> For those of you who are not just ready for work or are just
> getting up, get up lazy bones. It's time you got big day
> ahead. And the thought for today is, remember National
> Secretary's Week was last month.

CAPCOM Joe Engle signed off and handed over to Jack Lousma. It was a Sunday morning, and Stafford called out to Engle before he left the Control Center. Both men attended Seabrook United Methodist Church, less than two miles from Mission Control.[9] Stafford said, "How about doing me a favor, will you, ol' buddy?"

Engle said, "You name it."

"We're kind of out of town for church today, and the minister, Reverend [Bob] Parrot, wanted my reflections or something that might be appropriate to read in the service since I won't be around there. If you got a pencil, I copied down a couple of things."

"Roger. Go ahead."

"Psalm 8, Psalm 122, Psalm 148, and Isaiah 2:4. Tell the congregation hello for me."

"Roger that, Tom. That is very appropriate; I'll see that the word gets around."

The final telecast from Apollo 10 came five hours before entry. The crew took turns expressing their impressions of the flight.

> YOUNG: This is your old retired philosopher speaking to you
> from outer space and telling you that TV is on its way back.
> We have a little more work to do, and then we'll be back
> with you. And it will sure be great to be back. It's been

---

[9] Thank you to La-la Smith of Seabrook UMC for details about the church and minister.

utterly unbelievable, the mission has. We've really enjoyed every bit of it. So until we see you again, we'll say so long.

CERNAN: I can't tell you what a rewarding and satisfying experience this has been. It's had its moments, as I said. I'm just thankful that through the medium of television, we've been able to share it with so many people in real time. I'm convinced after this mission, none of them are going to be easy, but nothing is impossible. And I think the future of manned space flight for now and for many generations to come is going to uncover many, many other new challenges and experiences that we're really incapable of even conceiving at this time. It's been a great eight days. Of course, we're looking forward to getting home. And I guess next time we'll be talking to you and seeing you, we'll be back on the ground. Thank you.

STAFFORD: Good morning. On the final closeout telecast of Apollo 10, we just want to say that it has just been fantastic, the total views that we've seen on this total mission. Again, like Gene pointed out, no mission is easy, and it's been a lot of work, but we've enjoyed the whole thing greatly. And also, the main thing is we've been able to, in real time, on some of the major parts of the mission, to share this with you. Like we pointed out, that fantastic view when we left the Moon, man has certainly progressed a long way in such a short few years. And how much we're going to progress in the future is left to your imagination. But if we harness our energies and keep our perspectives right, the goals are unlimited.

Stafford changed the topic to the Lunar Module: "The ascent part of *Snoopy* is on its way around the Sun now. The descent part is still in an orbit around the Moon, so he's got quite a split personality." He continued, "*Snoopy* is a symbol of a manned flight awareness program and represents the good work and efforts of the hundreds of thousands of people who have made the manned space flight program so successful."

Stafford signed off, "And from the crew of Apollo 10, we'd just like to give all those people a salute and acknowledgment, and this is one way of doing it, just by naming a spacecraft after their symbol. And so from the five of us, Gene Cernan, John Young, Tom Stafford, *Snoopy*, and *Charlie Brown*, we'd just like to say goodbye."

Apollo 10 slammed into Earth's atmosphere traveling 36,360 feet per second (24,791 statute miles per hour), faster than any human traveled before or since. They splashed down in dawn's early light, three miles away from the recovery aircraft carrier, the *USS Princeton*. The successful flight of Apollo 10 set the stage for the first lunar landing.

Maskelyne, a milepost for the Apollo 11 descent, viewed from 50,000 feet. One of Snoopy's Reaction Control Thrusters is visible to the right. Photo credit: NASA AS10-29-4296.

John Young inspects Charlie Brown on the recovery carrier after the mission. Photo credit: NASA S69-20796, Kipp Teague, scan by J.L. Pickering.

14

WE SHOULD HAVE COME UP WITH A BETTER FAILURE

MISSION PLANNING

A strategic sequence of flights in Projects Mercury, Gemini, and Apollo taught NASA how to perform long-duration missions, rendezvous and docking in lunar orbit, and fly within 50,000 feet of the Moon's surface. But those last few miles presented a daunting challenge.

How do you land on the Moon in a spacecraft with no brakes, orbiting at one mile per second at an altitude of 50,000 feet? In general terms, you would tip the spacecraft on its back and use the descent engine to slow it down, firing its thrust against the direction of travel. As the forward velocity decreased, the spacecraft would drop out of orbit, gradually arcing toward the Moon. As the Lunar Module's trajectory transitioned from horizontal to vertical in relation to the Moon, so would its orientation, keeping its thrust pointed directly into the direction of travel. The last few hundred feet would see the LM drop straight down, balanced on its thrust until it softly settled onto the lunar surface. As the engine burned fuel, the spacecraft's weight would decrease, and so would the appropriate amount of thrust. These are some fundamental elements of a powered descent in an ideal scenario.

However, the Moon was far from an ideal environment. What if the astronauts did not like the landing area when the LM pitched over enough for them to see out their windows? They would use much precious fuel creating forward energy to find a new area. Therefore, a tradeoff would be necessary between the most economical use of fuel, as described above, and the capability for the astronauts to adjust their landing area while there was still enough altitude and forward momentum.

With these considerations in mind, the mission planners devised a three-phased strategy for the powered descent. The first was the Braking Phase, which lasted eight and a half minutes. The main objective for this segment was scrubbing orbital velocity in the most fuel-efficient manner. During that time, the LM slowed from 5,560 feet per second to 500 fps (roughly 3,300 to 300 nautical miles per hour), and its altitude decreased from 50,000 to 7,000 feet. The LM started the powered descent approximately 260 nautical miles east of the landing site. At the end of the Braking Phase, it was less than five nautical miles east of the target.[1]

The Braking Phase required several throttle changes. The descent engine began at 10 percent power while the computer determined the LM center of gravity, through which the thrust was fired. After 26 seconds, the engine throttled up to full capacity.

In testing, the Lunar Module descent engine showed excessive corrosion when the thrust level ranged from 65 to 93 percent. The flight planners avoided this by having the LM use full thrust until it needed less than 60 percent. About six and a half minutes into the Braking Phase, the LM engine throttled down to just under 60 percent and gradually reduced thrust as the LM weight decreased throughout the rest of the descent.[2]

The second step was the Approach Phase. It began with an early "pitchover" so the crew could see where the guidance system was targeting. The primary objective was to let the Commander assess the landing site while he still had enough altitude and forward momentum to make significant adjustments with minimal fuel usage. There was a tradeoff. The early pitchover resulted in less efficient use of propellant

---

[1] Floyd Bennett, "Apollo Experience Report: Mission Planning for Lunar Module Decent and Ascent," (Houston: Manned Spacecraft Center, 1972), 8–10, https://www.hq.nasa.gov/alsj/nasa-tnd-6846pt.1.pdf.
[2] Bennett, "Apollo Experience Report," 8–10.

since the thrust was now more vertical than the glide angle. Even with this concession, the pre-mission simulated landings occurred before the fuel quantity reached a critical level. The Approach Phase ended with the LM about 500 feet above the surface, less than half a mile east of the landing area.[3]

The third and final portion was the Landing Phase, which began with a 60 feet per second forward velocity and a descending rate of about 16 fps. The forward velocity was necessary because the Commander could not see anything below or behind the Lunar Module. When he found a suitable landing spot, he could pitch the LM back, using the thrust to nullify the forward rate. If all went well, the final hundred feet would be a gentle vertical drop balanced on top of the thrust.

PREPARATION FOR THE LANDING

The Simulation team began work on the first landing years before the world turned its attention to Apollo 11. Jay Honeycutt joined the Flight Control Division in February of 1966 and was assigned to the Simulation Branch, where Dick Koos was his supervisor and Harold Miller the Branch Chief. Many were already working on landing simulations by the time he arrived.[4] Honeycutt describes his team's role as the Lunar Module advanced from design phase to reality:

> We determined what the sim system should look like and developed a set of software requirements necessary to reflect what we thought the system should do. These requirements were passed on to the Flight Support division who had responsibility for programming and operating the IBM 360 computers dedicated to simulation.[5]

After the successful Apollo 7 and 8 flights in 1968, it was clear the first lunar landing attempt would occur the following year. Flight controllers gathered engineering data, analyzed the systems, broke down the schematic diagrams, and learned everything possible in their areas of responsibility. The controllers developed and refined "Mission

---

[3] Bennett, "Apollo Experience Report," 8–10.
[4] Jay Honeycutt, email message to author, April 6, 2020.
[5] Jay Honeycutt, email message to author, September 30, 2022.

Rules," which were guidelines for responses to any conceivable problem.

Each system and subsystem were discussed in fine detail. Meetings came out of the woodwork as flight controllers and their support teams rehashed Mission Rules, Mission Techniques, spacecraft changes, and software updates. Numerous small group meetings focused on the unending complexities of a lunar mission, therefore preparation for a space flight lasted over 18 months. For Apollo 11, the process was condensed to half that time, and every day was packed.

Simulators located at Cape Kennedy were essential for the flight crew preparation. The Lunar Module simulator was behind schedule because it had to reflect the constant changes to the actual Lunar Module and landing software. Armstrong and Aldrin, along with their backups Jim Lovell and Fred Haise, were finally able to use the simulator a few months before the flight. Three men shared the Command Module simulator. Mike Collins, the prime crew Command Module Pilot, had first priority, but his backups Bill Anders and Ken Mattingly also received time. Anders, the veteran of Apollo 8, was the original backup CMP. In early 1969, he accepted the role of Executive Secretary of the National Aeronautics and Space Council, advising the President and Vice President on space policies. He announced he would retire from the Astronaut office in August of 1969. If Apollo 11 were delayed, Mattingly needed to be ready.

"Anders already knew all he had to know, so they needed someone to fill in during the simulations," said Mattingly. He knew he had no real chance of flying, but to be involved on the backup crew was thrilling to Mattingly. "I thought I had died and gone to heaven this time."[6] For a young astronaut hoping to fly to the Moon, there were worse places to be than on the Apollo 11 backup crew with highly-respected astronauts Jim Lovell and Fred Haise.

While the flight crews had Command Module and Lunar Module simulators in which to practice, the flight controllers had their own simulations. Since the first Mercury flight, Chris Kraft firmly believed in having flight control sims to train for every conceivable scenario and

---

[6] Ken Mattingly, interview by Rebecca Wright for NASA Johnson Space Center Oral History Project, November 6, 2001, 28.

every possible failure. Some simulations focused on a small group of controllers, while others tested the whole team. There were also "integrated sims" where the astronauts participated in a simulator and worked with the Mission Control team.

The tight schedule before Apollo 11 not only affected the astronauts and flight controllers, but also impacted the instructors running the simulations. Jay Honeycutt, one of the Simulation Supervisors, looked up the statistics. "It turns out we did fewer integrated sims with the Apollo 11 crew than we did with any other mission. They had so many other obligations that they were not available for as many integrated sims as the crews for other missions were."[7]

Without normal access to the astronauts in training, Honeycutt utilized a valuable resource, Charlie Duke. Neil Armstrong selected Duke to be the Capsule Communicator (CAPCOM) for the lunar landing phase. Duke also represented the crew in the all-important Mission Techniques and Mission Rules reviews, relaying critical issues or decisions to the prime crew.

Duke also assisted Honeycutt at those meetings. "We would go to pretty much all of the meetings and make a point of sitting beside each other and decide if an issue was something we should run in a simulation, or if it was something for the flight crew to deal with, or something for the flight controllers," said Honeycutt.[8]

Honeycutt regarded Duke as "an honorary member of our team." The two worked well together. Honeycutt says, "Charlie Duke is my favorite of all the people that I had the opportunity to meet in project Apollo or otherwise, for that matter. I think the world of Charlie and have ever since the first time I had an opportunity to work with him. Incredible person. Not only a great fighter pilot and a great astronaut, but more importantly, a great person."[9]

Duke related well with Armstrong, Aldrin, and Collins and blended seamlessly with flight controllers on the previous mission. "We felt like he was one of our own," said Flight Director Gerry Griffin.[10] He was the

---

[7] Jay Honeycutt, interview by author, March 26, 2019.
[8] Jay Honeycutt, interview by author, March 26, 2019.
[9] Jay Honeycutt, interview by author, March 26, 2019.
Jay Honeycutt, interview by author, October 4, 2022.
[10] Gerry Griffin, interview by author, March 19, 2019.

ideal person to be the interface between Mission Control and the astronauts during the intense landing phase.

SIMULATIONS BEGIN

Each of the controllers developed and refined their part of the Mission Rules, and then they began some standalone simulations amongst their team and their support staff. The astronauts were not involved at this point.[11] The number of participants in the sims would grow and eventually include the Manned Space Flight Network, maintained by the Goddard Space Center in Maryland. Goddard was responsible for overseeing the massive network of 29 tracking stations, four ships, and eight aircraft, all delivering vital communications to Mission Control.[12]

These were called Simulated Network Simulations, or "Sim-Net-Sims," providing training not only for the flight controllers but also the ground controllers who made Mission Control operations run without a glitch.[13] The Sim-Net-Sims were the penultimate level of simulation, topped only by the "Integrated Sims," the crown jewel.[14]

Integrated Sims involved the astronauts, flight controllers, support staff, contractors, the network, the real-time computer complex, and everyone who might be called on to provide support during the actual mission. The simulators and consoles contained the latest software from MIT. They were extremely realistic rehearsals.[15]

The trainers running the simulations were led by Simulation Supervisors called Simsups. Kranz described their role:

> The Simsup's job was to come up with mission scenarios that were utterly realistic and trained every aspect of the crew, controllers, and the Flight Director's knowledge. They tested every aspect of the procedures and planning that we had put together. They tested our ability to innovate strategies when

[11] Doug Ward, interview by author, March 16, 2019.
[12] NASA, "The Manned Space Flight Network for Apollo," (Greenbelt, Maryland: Goddard Space Flight Center), 1968,
https://web.mit.edu/digitalapollo/Documents/Chapter8/apollonsfn.pdf.
[13] Francis E. "Frank" Hughes, interview by Rebecca Wright for NASA Johnson Space Center Oral History Project, September 10, 2013, 10–11.
[14] Doug Ward, interview by author, March 16, 2019.
[15] Doug Ward, interview by author, March 16, 2019.

things started to go bad. And training in Apollo was realistic…
you would get sweaty palms, and when the pressure was on in
a training episode, it no longer was training. It was real; the
same emotions, the same feelings, the same energies, the same
adrenaline would flow.[16]

Astronaut Fred Haise had another description of Simsups, "A group
of rather devious people that had been assembled to devise the skits for
these simulation scenarios to put in specific failures at specific times."[17]
The simulations were so intense the participants consistently said they
could not tell the difference between the practice runs and the actual
flights. Controller John Aaron said, "When you walked out of the room
at the end of the day, you were drained. I used to tell people if you could
survive the simulations, the mission is a piece of cake. Because you're
not usually working 20 problems at once."[18]

College football teams have scout teams mimicking the opposition
so the starters will have a good picture of what they will encounter in
the actual game. Imagine a scout team full of All-Stars; that is what the
flight controllers faced in training. Those All-Stars on the simulation
team lurked in their room to the right of the Mission Operations Control
Room (MOCR), separated from their prey by a window through which
they viewed the flight controllers. They were a formidable team.

Kranz described them as a dedicated team of trainers "with the
mandate to find the holes, exploit personnel issues, challenge team
integrity, and validate the planning. They closely monitored the Mission
Rules discussions and exploited any ambiguities or inconsistencies in
our own relations or in the relationships we had with the crew."[19]

Forging the astronauts and flight controllers into one team was vital
to the space program's success, but it was difficult. The astronaut office
consisted of fighter pilots and test pilots with advanced engineering

----

[16] BBC "13 Minutes to the Moon," Season 1, Episode 02, "Kids in Control,"
presented by Kevin Fong, aired May 22, 2019,
https://www.bbc.co.uk/programmes/w3csz4dk.
[17] Fred W. Haise, Jr., interview by Doug Ward for NASA Johnson Space Center Oral
History Project, March 23, 1999, 19.
[18] "13 Minutes to the Moon," Season 1, Ep 02.
[19] Gene Kranz, 2018 Vice Adm. Donald D. Engen Flight Jacket Night Lecture.
National Air and Space Society, Washington, DC, November 8, 2018,
https://airandspace.si.edu/events/flight-jacket-night-eugene-kranz.

degrees. These men were experienced, intelligent, skilled, highly trained, used to leading, and had loads of confidence. The grueling astronaut selection process eliminated all but the cream of the crop.

The young flight controllers, typically a few years removed from college, were not pushovers either. Simply put, if you did not have a high degree of self-confidence, you did not survive long in Mission Control. Intense training and the pressure of making split-second, life-or-death decisions weeded out all who did not thrive in such an environment. Jay Honeycutt puts it in perspective:

> It's an interesting phenomenon. Some people love to be in a position of responsibility. They love to have to make decisions. Others can't handle it. So when you put a flight control team together, you get the guys who want to be in charge, and they want to have the responsibility, and they want to have to make the decisions in real-time, and they want to be able to prove that you cannot throw anything to them that they cannot handle. That is the type of personality that is necessary to do the job.[20]

The simulation team not only prepared the controllers and astronauts for all possible scenarios, but also caused the two forces to work together solving problems. Honeycutt said, "Our job was to meld them into a team that had confidence in the other's ability to do whatever the issue might be and to acknowledge that the other had more data and more time and more resources to come up with the decision."[21]

The trainers watched the young men in the MOCR and listened to their conversations with each other and their back rooms, occasionally hearing some unflattering comments. The simulation team monitored the amount of traffic on the Flight Director's loop. If a controller did not have much to do, the trainers took it as a personal affront and quickly remedied the situation.[22]

The simulations were demanding. The dreaded debriefings afterward were even worse. Jay Honeycutt explains:

---

[20] Jay Honeycutt, interview by author, March 26, 2019.
[21] Jay Honeycutt, interview by author, March 26, 2019.
[22] Jay Honeycutt, interview by author, March 26, 2019.

The Flight Director would talk to each one of his console positions. Whichever ones happened to be the recipients of our attention had to explain how they handled it, why they did it, and if there were better options. Then we were the last to speak. We would give them our opinions on where they could have done this this way, or could have done that that way, and might have had a little better result.[23]

The simulation team not only evaluated the decisions made by the controllers but also how long it took to make the call and how they did so. The first simulation Jay Honeycutt led involved a series of launch aborts due to malfunctions in the Saturn V. There were different abort procedures, or modes, depending on how far into the flight the problems occurred. It was a complicated process to determine which mode was correct. The Retrofire Officer was John Llewellyn. Honeycutt describes the scene:

John was an icon. He served as a Marine in Korea, one percent body fat, all Marine. He personified a tough guy. He was a great guy, but he had that look about him. In those days, aborts would either go in the Atlantic, or in Africa some place, or you would abort to orbit. And John had a bunch of back-of-the-envelope calculations he had done. So as soon as the engines cut off, he would fire back to the Flight Director what mode of abort based on the time the engines failed. And then he would query the mission computer to get the absolutely accurate answer. If it were different than what he originally said, he'd call up a change.[24]

In this simulation, Honeycutt caused the engines to fail as the abort mode changed from one to another. He also depressed the trajectory, so the abort mode decision was not simply dependent on the time of the failure. For good measure, he threw in a malfunction on the main computer in the Control Center. John got caught on the line between one mode and the next. So, he passed up his information from the back of his envelope, which was wrong. He couldn't get the right solution

[23] Jay Honeycutt, interview by author, March 26, 2019.
[24] Jay Honeycutt, interview by author, March 26, 2019.

from the computer, and the crew crashed."[25] Honeycutt was not looking forward to encountering Llewellyn afterward:

> We ran three or four more cases, and then the day was over, and John is still sitting at his console. I got up to leave and I had to go right by him, and I wasn't sure whether he was going to jump me. I go walking by and he calls me over there, and I thought, "Well, here it comes." And he says, "Thank you. I've been doing that trick for two years around here, and I never got caught. And you just proved to me that it was the dumbest thing I could possibly do, so I really appreciate what you did." And from then on, we were fast friends.[26]

"The simulations were the key to the program," said Gerry Griffin. "The training was rigorous. We ran hundreds of simulations and made our mistakes there instead of in flight."[27]

PREPARATIONS FOR APOLLO 11

The integrated simulations for Apollo 11 began in April 1969, three months before the launch. The astronauts were in the Cape's Command Module and Lunar Module simulators since they were updated with the latest flight hardware and software versions. As with any flight, there were plenty of launch abort sims since it was one of the most dangerous phases of the mission. But many simulations were focused on the mission's new aspect, the lunar landing. The trainers gave special "opportunities" to the flight controllers who had to make abort decisions during the descent.

One of those recipients was Steve Bales, the Guidance Officer (GUIDO) for the Apollo 11 landing.[28] Bales described his role recently to a BBC audio crew for their highly recommended series of podcasts titled *13 Minutes to the Moon*:

[25] Jay Honeycutt, interview by author, March 26, 2019.
[26] Jay Honeycutt, interview by author, March 26, 2019.
[27] Gerry Griffin, interview by author, July 13, 2020.
[28] On the communication loops, his call sign was "Guidance" since "Guido" was difficult to distinguish from "Fido," the Flight Dynamics Officer.

There were six Guidance Officers, but we all specialized in various parts of the Apollo missions. I specialized in ascent and descent. That means I had to learn how to take the landing computer, which was not powered up until the day of descent, get it powered up, send information to it so it knew where it was and where it was going to be at landing, and then monitor it during the landing to make sure it was doing the right things, because for the first two-thirds of the landing, they were on autopilot. That vehicle wouldn't go to the Moon if that computer wasn't working and they knew it, so my job was to make sure the computer and its sensors were working properly. That's the Guidance Officer's job.[29]

When preparations for the lunar landing began, controllers initially thought the Flight Dynamics Officer (FDO) would make most of the abort decisions. However, it turned out the Guidance Officer was monitoring the systems at a level that he would be the first to detect problems requiring an abort.[30]

It also became apparent if there was an abort during the descent, the requirements for the ensuing rendezvous with the Command Module would be tighter than the limits for landing. A realistic scenario emerged where the descent could be aborted due to rendezvous concerns even when a successful landing was still possible. FDO Jay Greene explained:

As you're going down, the first thing we monitored for was to keep the crew from crashing into the Moon. That would have ruined everything. But the other thing we monitored for was maintaining the ability to leave the trajectory we were on and make it back to rendezvous with the Command/Service Module. And if we ever lost that rendezvous capability, or prior to losing that capability, we would abort the descent. So, the FDO's job evolved into one of monitoring this trajectory and always keeping a rendezvous capability so that anytime I aborted, I had a plan for how to get back to the CSM quickly. That turned out to be a full-time job.[31]

---

[29] "13 Minutes to the Moon," Season 1, Ep 02.
[30] Steve Bales, email message to author, March 2, 2020.
[31] Jay H. Greene, interview by Sandra Johnson for NASA Johnson Space Center Oral History Project, November 10, 2004, 18.

There were two types of simulations for the landing phase. The first was called Descent Preparation. "They would take almost the whole day," said Bales. "You'd start out in the morning and do a total dress rehearsal of everything you had to do from Lunar Module power-up through the four revolutions before landing through the actual landing. That was hard to do because the simulator didn't always stay up."[32]

The other simulation type was known as Descent/Descent Aborts. They focused on the time right before, during, and after the actual descent and lasted about 20 minutes each. As the controllers debriefed afterward, the simulators were set up for the next run. They could do six or seven runs each day unless there was an issue with the simulator.[33] The Simsups would usually target the people who had abort calls. Bales received his share of problems.

The schedule was jam-packed before each flight. There were simulations for all phases of the mission, especially launches, launch aborts, lunar insertion, descent preparation, descent, and ascent from the lunar surface. Flight controllers spent the rest of their time going to meetings and writing Mission Rules.[34]

Bales remembers the first simulation of the Apollo 11 powered descent, which occurred before the launch of Apollo 10. Anytime they ran the initial simulation of its kind, for example, the first launch of a manned Saturn V on Apollo 8, the tension was intense. Since this exercise involved landing on the Moon, many leaders within NASA, including Chris Kraft, gathered in the Control Center to watch. The presence of the managers meant the simulation was the most important thing they could be doing that day. Their arrival did not go unnoticed by the flight controllers. Jay Greene told Steve Bales, "This is exactly like the real deal."[35]

Chris Kraft was not content experiencing the first descent simulation from his customary spot in the rear of Mission Control; he wanted to observe from where the action was. Each console had a place for the flight controller and his assistant, with two communication jacks on each side. Bales was caught off guard when Kraft plugged in a

[32] Steve Bales, interview by author, October 9, 2019.
[33] Steve Bales, interview by author, October 9, 2019.
[34] Steve Bales, interview by author, October 9, 2019.
[35] Steve Bales, email message to author, August 29, 2020.

headset and sat beside him on the front row. The tension ratcheted up even more.

The simulation team intended the first sim to be nominal, with no malfunctions, so the astronauts and flight controllers could work together for a successful landing. Unfortunately, the simulator was configured improperly, resulting in a misalignment of the LM guidance platform. At the beginning of the powered descent, the computer thought it was flying roughly parallel to the lunar surface when it was actually descending. "By the way," notes Bales, "the same effect happens when the vehicle believes it is in the correct location at powered descent ignition but indeed is four miles downrange—this was the actual Apollo 11 experience."[36]

As the first simulation progressed, Bales quickly noticed the descent velocity was greater than expected. If the error reached 35 feet per second, Mission Rules required an abort. Beyond that rate, the Primary Navigation and Guidance System (PNGS, pronounced "pings") could not have safely aborted, even though it might have been able to continue the descent to an altitude where the astronauts could assume manual control. It was one of the scenarios where an abort would be caused by rendezvous considerations even though a successful landing was still possible. "To say the least, this would have been difficult to explain to the general public," said Bales.[37]

Around four minutes into the sim, the error crossed the 35 feet per second limit, and Bales called an abort. Flight Director Gene Kranz gave the official order, which CAPCOM radioed to the crew. They used the descent engine to begin the rendezvous with the Command Module. "The abort always began on the descent stage [using the descent engine] unless that engine had failed, or the vehicle was so low that the descent fuel was almost depleted," Bales points out. The descent engine was more powerful than the ascent engine, and utilizing it after an abort avoided a staging maneuver during an unsettled part of the flight.[38]

Bales knew the problem had all the symptoms of a misaligned guidance platform. The Lunar Module gained altitude and headed toward a rendezvous with the Command Module. Since the day was

[36] Steve Bales, email message to author, August 26, 2020.
[37] Steve Bales, email message to author, August 26, 2020.
[38] Steve Bales, email message to author, August 26, 2020.

devoted to Descents and Descent Aborts, they continued the simulation until the LM reached orbit. Bales requested that the sim carry on for about five extra minutes so Neil Armstrong could check the platform alignment. Armstrong reported a misalignment of 0.5 degrees, a significant error that eventually would have required an abort. Bales had made the right call. After hearing the report, Kraft slapped him on the shoulder, said, "Good job," and returned to his seat on the back row.[39]

The sim team performed a debriefing while the equipment was prepared for the next run and acknowledged it was an unintended simulator problem. "They thought it had been corrected," said Bales, "but the same thing showed up on the second run, which I also had to abort. Then Gene told them to fix it or we would stop the sessions for the day." It took some time, but they finally solved the problem.[40]

At first, there were many crashes as the flight controllers got used to landing on the Moon. They added Mission Rules to handle unforeseen failures thrown in by the sim team. The simulated communication delay from the Moon to Earth added to the complexity. Radio waves traveling at the speed of light take about a second and a half to span the quarter of a million miles separating the two celestial bodies. The flight crew often found themselves out of synch with Houston due to the response time and delay from signals going back and forth.

The hectic training schedule took a toll on Armstrong, Collins, and Aldrin. They trained all week in the simulators at the Cape for up to 14 hours a day, then flew home to see their families for a precious few hours before heading back to Florida. Janet Armstrong was concerned about her husband and his crew:

> Neil used to come home with his face drawn white, and I was worried about him. I was worried about all of them. The worst period was in early June. Their morale was down. They were worried about whether there was time enough for them to learn the things they had to learn, to do the things they had to do if this mission was to work.[41]

---

[39] Steve Bales, email message to author, March 2, 2020.
[40] Steve Bales, email message to author, August 28, 2020.
[41] Neil Armstrong, Edwin E. Aldrin, Jr., and Michael Collins, *First on the Moon: A Voyage with Neil Armstrong, Michael Collins, Edwin E. Aldrin, Jr*, with Gene Farmer and Dora Jane Hamblin, (Boston: Little, Brown, and Company, 1970), 47.

GENERAL SAM PHILLIPS

It was time to decide if the mission could launch on its scheduled date of July 16, 1969. If so, the ground crew needed to start preparing the Saturn V, which was already on the launch pad. If the verdict were to postpone, the mission would have to wait a month before the lighting conditions on the Moon were ideal again for a landing attempt. The burden fell into the lap of General Sam Phillips, the Director of the Apollo Manned Lunar Landing Program. His pre-NASA experience directing the Air Force Minuteman Intercontinental Ballistic Missile Program and the important decisions he made in the heat of the 1962 Cuban Missile Crisis made him the perfect choice to lead the Apollo program in times like this. John Noble Wilford summarizes, "Phillips was to emerge as the key decision maker, prodding companies to get their work done on time, monitoring their costs and performance, giving the go-ahead signal for flights."[42]

Phillips said, "I'm at the level which knows all the things you have to know to make a major decision. Below the program director, there isn't anyone who has the whole picture. Above the program director, the men have so many other responsibilities."[43] Now it was up to Phillips to decide whether to press on for the July 16 launch.

On June 12, Phillips organized a high-level conference call from his office in Washington, DC. Joining from Houston were George Low, Chris Kraft, Gene Kranz, and Dr. Charles Berry, Director of Medical Research and Operations. Deke Slayton was on the line at Cape Kennedy. The General began by saying, "I am fully ready to delay if something is not ready, or if we are pushing these men too hard. If we are doing that, then we will reschedule for August."[44] He listened to the various opinions, then asked Slayton. Deke acknowledged the fatigue level of the astronauts but said they were past the worst part of the training, and he recommended proceeding with the July launch. At the end of the phone call, Phillips made the final decision that impacted

[42] Wilford, *We Reach the Moon*, 64.
[43] Wilford, *We Reach the Moon*, 228.
[44] Armstrong, Collins, Aldrin, *First on the Moon*, 49.

history. He said, "Go."[45] Janet Armstrong immediately discerned a positive change in her husband's demeanor:

> The turning point came with the decision to go. After that, everything seemed to get better. They knew they were going, and the knowledge seemed to take the weight off their shoulders. That last weekend they were home, they were in great shape. When I talked to Neil on Monday [two days before launch], he was ready to go. When I talked to him on Tuesday, he was *really* ready. They were ready to go *then*.[46]

FINAL SIMULATION

The rigorous training bore fruit as the time approached for the Apollo 11 mission, and flight controllers experienced more and more successful landings. The final day of simulations occurred on July 5, 1969, 11 days before launch. Customarily, the trainers would take it easy on the troops, allowing the last simulation to end well and giving the controllers confidence going into the actual mission. Not this time.

In the months leading up to Apollo 11, one of the simulation supervisors, possibly Dick Koos, picked the brain of Jack Garman about some potential failures to include in simulations. Garman, a 24-year-old computer expert, was on the support team for Steve Bales. He happened to mention some alarms deep inside the bowels of the Lunar Module computer software. They were not meant to occur during the mission but were debugging tools for software programmers. It did not stop Koos from inserting the scenario into the final sim.

Steve Bales settled into his chair at the Guidance console, never imagining how vital the day would be. Commander Neil Armstrong and Lunar Module Pilot Buzz Aldrin were not even in the Lunar Module simulator; it was manned by the Apollo 12 backup crew of Dave Scott and Jim Irwin.[47]

A few minutes into the final simulation, Dick Koos dropped the bomb. The astronauts in the LM simulator were startled as an alarm

---

[45] Armstrong, Collins, Aldrin, *First on the Moon*, 50–51.
[46] Armstrong, Collins, Aldrin, *First on the Moon*, 51.
[47] Kranz, *Failure Is Not an Option*, 268. Houston, *Go Flight*, 154.

blared in their ears. It was one of the computer alarms Garman casually mentioned months earlier.

All eyes shot toward Steve Bales, who had no idea what was happening. In his many hours of study and preparation, he devised a response to every conceivable issue which could arise during the descent—but computer alarms were not on anyone's list. Bales remembers the alarm being a 1210, meaning multiple people were trying to use the computer simultaneously.[48] Regardless of the cause, he had to decide whether to abort or proceed. The alarm signaled a computer overload. Since the computer was flying the spacecraft on autopilot for most of the descent, Bales called for an abort.

Then it was time for the debriefing. Kranz was upset at Koos for throwing a curve ball in the final sim and using a failure that could not occur during an actual mission. The disagreement escalated as Koos told Bales he should have kept going with the descent. Koos pointed out the computer was still performing all of its primary requirements despite the alarm. He also reminded Bales that two cues were needed for an abort. Bales made his decision based on only one. The spacecraft's trajectory and velocity were still fine, so he should have continued.

After the debriefing, Kranz told Bales he wanted a list of all the possible computer alarms hidden in the software and a set of Mission Rules describing how to respond to each. Bales already had a million things to do in preparation for the mission, but that excuse didn't fly with Kranz. Bales agreed to do it even though he thought it was a waste of time.[49]

He delegated the task to Garman, who had been in the back room during the episode. Removed from the intensity inside the MOCR, the support staff were not as fired up about the situation. Ironically, Garman considered the program alarm "screwy" and thought, "we should have come up with a better failure."[50]

Steve Bales and Jack Garman had their marching orders. Garman contacted the experts at MIT who developed the computer and software. "MIT is the technical center of the universe, the largest collection of

[48] Steve Bales, interview by author, October 9, 2019.
[49] Steve Bales, interview by author, October 9, 2019.
[50] John R. Garman, interview by Kevin M. Rusnak for NASA Johnson Space Center Oral History Project, March 27, 2001, 21.

super-active brain cells on the planet," says Dr. Duncan Rankin, who did postgraduate work at MIT. "Nothing compares to it, certainly not the privileged little school down the street named Harvard."[51]

Garman asked for a list of all alarms and the appropriate response for each if encountered during the mission. The computer experts thought he was crazy, noting many of those alarms were simply debugging tools for programmers. Garman knew Kranz demanded an answer, so he persisted until MIT gave him all the alarms and how to handle each one.[52]

The LM computer processed the guidance equation every two seconds, incorporating new navigation data and refining the trajectory.[53] The software prioritized the necessary tasks and was programmed with a safety feature. If the computer were overloaded and unable to accomplish all of its functions within the two-second time frame, it would reboot and set off an alarm like the one in the sim. The computer would immediately restart, dump the lower priority items, and carry on with the higher priority tasks such as maintaining the correct trajectory, velocity, and attitude.

When Garman received the complete list from MIT, he met with Bales to discuss which alarms required an abort and which meant they could continue. Garman created a hand-written chart that would become one of the most important cheat sheets in history. He gave a copy to Bales and his boss Jerry Bostick, Chief of the Flight Dynamics Branch.[54] Bales had to sign it before the launch to show he had reviewed the list.[55] They would place a copy under the glass on their consoles before the actual descent. The flight controllers were finished training. They knew they were ready.

[51] Duncan Rankin, email message to author, November 11, 2019.
[52] Steve Bales, interview by author, October 9, 2019.
[53] Don Eyles, *Sunburst and Luminary: An Apollo Memoir*, (Boston: Fort Point Press, 2019), 24.
[54] Jerry Bostick, interview by author, August 25, 2020.
[55] Steve Bales, interview by author, October 9, 2019.

*Christopher Kraft developed the concepts of Mission Control
and flight simulations. Photo credit: NASA.*

*Steve Bales manning the GUIDO console
on the front row of Mission Control. Photo credit: NASA.*

Escape Tower

Command Module

Service Module

SLA Adapter

White Room

Third Stage

Apollo 11 in front of the Launch Umbilical Tower.
Astronauts and technicians rode an elevator to the 320-foot level and
walked across Swing Arm 9 to access the White Room and Command Module.
Note the technician visible on top right of the White Room. The Lunar Module
was tucked inside the cone-shaped Spacecraft-Lunar Module Adapter (SLA).
Below the adapter was the black Instrument Unit and third stage
of the Saturn V launch vehicle. Photo credit: NASA AS11-69-717,
Kipp Teague, annotations by author.

15

YOU'LL NEVER BEAT OUT THE THUNDERBIRDS

JULY 16, 1969

The launch of Apollo 11 was a global event. Worldwide media flocked to Cape Kennedy, including Derryn Hinch of the Macquarie Broadcasting Network, a commercial radio network in Australia.[1] While Walter Cronkite described the historic event with a sense of wonder, Hinch's excited commentary was, in the words of Colin Mackellar, "like a race call."[2]

> HINCH: It's one minute and one second, and we're going into the final countdown on man's flight to the Moon. And it's just absolutely incredible here. The rocket is just *raring* to go, it's 50 seconds now until man takes off on this flight, and frankly, I just can hardly describe to you the excitement and the tension. Well, this time we're really going, and it starts a whole new phase of manned flight.
>
> Down to 30 seconds, we've got so little time now before we finally take off. It will lift off very slowly, of course, ignition will come at eight seconds probably and there are

---

[1] Colin Mackellar, email message to author, April 22, 2021. Hinch parlayed his media popularity and larger-than-life personality into a position in federal politics. He displayed his interest in the space program with his appearance at Honeysuckle Creek's 50th Anniversary Celebration of Apollo 11.

[2] Colin Mackellar, email message to author, July 21, 2020. Colin maintains the website honoring the Honeysuckle Creek Tracking Station, https://www.honeysucklecreek.net/.

these big sort of brake things which will hold it down until it gets to zero, and then it will take off very slowly, so slowly you'll think it's going to fall over. Now I'll give you the final countdown. Ten. Nine. Eight. Seven. Six, we have ignition sequence. Five. Four. Three. Two. One. Zero.

And the flames, you can see them now, and she's ready to go, it's starting to move. You can hear the noise. You can probably hardly hear me, but there she goes, and it's pulling away, and the crowds are yelling, "GO, GO, GO, you beaut!" And now you hear the rumble, the stands where I'm sitting are shaking, the lights and the stands are shaking. And there it goes, it's incredible, with huge white flames. There is smoke all around. I'll let you listen for a second to the noise.

The people are on their feet now saying, "GO YOU BEAUTY, GO!" And here it goes now with a roar, we can still see it, and it's going into a very small patch of cloud, it's going to become clear again now. There we go, it's firing, it's an absolute beauty. You can still hear that sound. It's the greatest thing anybody is ever going to see. I know that as long as I live, I am never going to see anything quite like this one. The Americans have really done it, I have to tell you that.[3]

Charlie Duke arrived in Houston from Cape Kennedy and assumed his duties as CAPCOM six hours into the mission.

DUKE: Hello, Apollo 11. Houston. Be advised your friendly White Team has come on for its first shift. If we can be of service, don't hesitate to call.
COLLINS: Thank you very much.

PUBLIC AFFAIRS OFFICER (DOUG WARD): This is Apollo Control at six hours and thirty-one minutes. At the present time, the spacecraft is 29,363 nautical miles from Earth, and the velocity continuing to drop off gradually, reading now 11,192 feet per second. Gene Kranz has taken over as Flight Director from Clifford Charlesworth. Kranz has been

---

[3] For more on Derryn Hinch and his Apollo 11 launch audio, *see* www.honeysucklecreek.net/msfn_missions/Apollo_11_mission/a11_launch_Hinch.ht ml.

*reviewing the status of the spacecraft's systems with his team
of flight controllers; everything looks very good at this point.*

The following morning, Apollo 11 backup Commander Jim Lovell
took over as CAPCOM for a few minutes to have some fun with Neil
Armstrong.

> **LOVELL:** Is the Commander aboard?
> **ARMSTRONG:** Roger, the Commander is aboard.
> **LOVELL:** I was a little worried. This is the backup
> Commander still standing by. You haven't given me the
> word yet. Are you Go?
> **ARMSTRONG:** You've lost your chance to take this one, Jim.
> **LOVELL:** Okay. I concede.

Gene Kranz's White Team came on shift again in the middle of the
afternoon. The highlight of the day would be a 6:30 p.m. live television
broadcast. The crew turned their camera on early so technicians at the
Goldstone tracking station could ensure everything was working
properly. They were filming scenes inside the cabin.

> **COLLINS:** Boy, you sure get a different perspective in this
> thing in zero G. Right now, Neil's got his feet on the forward
> hatch, and he can, with his arms, reach all five windows. He
> can reach down into the Lower Equipment Bay where the
> overboard drain is. He can practically reach over in the
> cockpit.
> **DUKE** (LAUGHING): Sounds like Plastic Man to me.
> **COLLINS:** I'm hiding under the left-hand couch, trying to stay
> out of his way.
> **DUKE:** A good idea, Mike.

Buzz Aldrin asked about the status of Luna 15, an unmanned Soviet
spacecraft launched a few days before Apollo 11. The Russians hoped
Luna 15 could land on the Moon and return samples of the lunar soil
before Apollo 11 splashed down, thus taking some of the luster out of
the American mission. Duke said, "Latest on Luna 15: TASS [the Soviet
Press Agency] reported this morning that the spacecraft was placed in
orbit close to the lunar surface, and everything seems to be functioning

normally on the vehicle. Sir Bernard Lovell says the craft appears to be in an orbit of about 62 nautical miles."

Duke told Collins, "President Nixon has declared a day of participation on Monday for all federal employees to enable everybody to follow your activities on the surface. Many state and city governments and businesses throughout the country are also giving their employees the day off. It looks like you're going to have a pretty large audience for the EVA."

Collins said, "Oh, that's very nice, Charlie. I'll tell them about it."

At 6:30 p.m., the television show began with the distant Earth out the window. Armstrong began, "Apollo 11 calling in from 130,000 miles out. And we'll zoom our camera in slowly to get the most magnification we can." He described the visible land masses and cloud patterns.

Collins jumped on the communications loop before rolling the camera in a full circle. "Okay, world, hold on to your hat. I'm going to turn you upside-down."

Duke said, "11, that's a pretty good roll, there."

"Oh, that was pretty sloppy, Charlie. Let me try that one again."

"You'll never beat out the Thunderbirds," said Duke.[4]

They turned the camera inside the cabin and showed each of the astronauts. Collins said, "It looks like it's probably your dinner time down there, Earth. We'll show you our food cabinet here in a second." He demonstrated the food storage compartment and described how they used a water gun to make chicken stew.

Near the end of the broadcast, Duke told the crew how impressive the clarity and details were on the TV.

Collins replied, "You can take a look at all the circuit breakers; make sure the right ones are in and the right ones are out."

"Roger. Big Brother's watching."

"And we're glad of it. Boy, you guys have sure been doing a good job of watching us, Charlie. We appreciate it."

The crew signed off slightly after 7:00 p.m.

---

[4] The Thunderbirds are the US Air Force Demonstration Squadron.

PAO (*WARD*): *At this time, Apollo 11 is 137,219 nautical miles from Earth. The spacecraft is traveling at a speed of 4,132 feet per second. Here in Mission Control, things have settled down into a rather quiet nighttime routine.*

Neil Armstrong walks toward the White Room before launch. Photo credit: NASA.

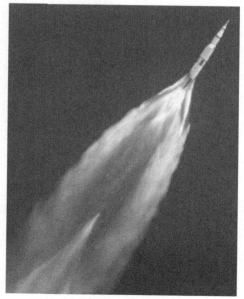

Apollo 11 photographed from a KC-135 Apollo Range Instrumentation Aircraft (ARIA) from an altitude of 43,000 feet. Photo credit: NASA.

Apollo 11 Launch.
Photo credit: NASA KSC 69PC-442.

16

IF THAT'S NOT THE EARTH, WE'RE IN TROUBLE

SECOND TV BROADCAST, FRIDAY, JULY 18, 1969

Cliff Charlesworth's Green Team of flight controllers replaced the Black Team early Friday morning. Bruce McCandless was the new CAPCOM, taking over from Ron Evans, who manned the post during the overnight shift. The crew woke up a little after 8:30 a.m., and it did not take long for Mike Collins to continue the good-natured banter with Mission Control.

> COLLINS: How's the old Green Team this morning? Did you have a quiet night?
> McCANDLESS: Yeah. It was a very quiet night down here. The old Black Team is complaining they didn't get a chance to make any transmissions.
> COLLINS: Well, we'll be seeing them tomorrow, I guess.
> McCANDLESS: Yeah. Ron's getting to be known as the silent CAPCOM.
> COLLINS: Those are the best kind, Bruce!

Buzz Aldrin described the weather on Earth from 160,000 nautical miles away. He also noted, "The eastern Mediterranean is phenomenally clear. You can see all the lakes; the Dead Sea stood out quite well."

McCandless was curious about the limit of resolution for the 28-power sextant. He asked Aldrin what was the smallest object he could see from that distance.

"Well, you can see the Nile River going almost up to its source. The lake is obscured by clouds, but you can trace it all the way on up… I guess that's down, though, isn't it?"

Apollo 11 wives Janet Armstrong, Pat Collins, and Joan Aldrin gathered around the Aldrin pool Friday afternoon while their children swam and played. They relaxed together, waiting to see their husbands on a television broadcast scheduled for later that afternoon. Armstrong and Aldrin were going to check out the Lunar Module and transmit from inside *Eagle*. Janet and Pat left for their homes to get ready for the telecast. Mrs. Armstrong turned on her car radio and was surprised to hear the show had already begun.[1]

The wives were not the only ones caught off guard. Gene Kranz's White Team had relieved the Green Team earlier in the afternoon. The network technicians in Mission Control announced the astronauts had turned on the television camera as they made their way into the Lunar Module. CAPCOM Charlie Duke let the crew know the Goldstone tracking station was receiving live TV. Aldrin said, "this is just for free," before the scheduled broadcast began.

> **PUBLIC AFFAIRS OFFICER** (DOUG WARD): *Getting a very good view of the work going on in the Command/Service Module tunnel. That appears to be Neil Armstrong working in the tunnel working on the drogue and probe assembly. This extremely sharp, clear picture is coming to us from about 175,000 miles distance from Earth, presently about 48,000 miles from the Moon.*

There was not much light in the tunnel, but the TV camera could make out the details. As Armstrong inspected the probe, he commented, "Mike must have done a smooth job in that docking. There isn't a dent or a mark on the probe."

Duke said, "We see lots of arms."

---

[1] Armstrong, Collins, Aldrin, *First on The Moon*, 185.

Collins replied, "The only problem, Charlie, is these TV stagehands don't know where to stand."

"Well, he doesn't really have a union card. We can't complain too much, I guess."

Television coverage of the missions seems like a foregone conclusion to the modern reader, but Chris Kraft and others had to push for its inclusion in the Apollo program. Doug Ward said, "Chris Kraft was probably more instrumental in getting television on the spacecraft and having television part of the Apollo missions than any single individual."[2] Kraft told his flight operations team he wanted to make sure the television worked well on Apollo 11. Ward added, "Jim Satterfield was the person handpicked by Chris Kraft to assure the Apollo 11 lunar TV was received and distributed successfully."[3]

Satterfield set up a television monitor on the back row of Mission Control, to the left of the Public Affairs Officer console. The monitor was rotated toward the front of the large room, making it viewable for Charlie Duke as he talked to the crew from the second row. During the TV broadcast, Duke casually sat on a console workstation facing backward toward the monitor.[4] This was in sharp contrast to the posture Duke would assume two days later, hunkering down behind the CAPCOM console studying data screens and punching communication loops during the tense final descent to the lunar surface.

In the spacecraft, Mike Collins said, "Well, the docking latches look good today, just like they did yesterday. Everything up in there looks just fine."

> DUKE: 11, Houston. We can even read the decals up there on the LM hatch.
> ALDRIN: Well, let me zoom it up and see how much you can read.
> DUKE: We see you zooming in on one of the decals now: "To reset, unlatch handle, latch behind grip, and pull back two full strokes." That's about all we can make out.
> ALDRIN: Hey, you get an A plus.

---

[2] Doug Ward, interview by author, March 16, 2019.
[3] Doug Ward, email message to author, March 17, 2019.
[4] Thank you very much to video archivist Stephen Slater for sending a video of this scene.

DUKE: Thank you very much, sir. At least I passed my eye test.[5]

COLLINS: I'm standing six feet from it, Charlie, and you can read it better than I can. There's something wrong with this system.

Armstrong and Aldrin were ready to go into the LM. As they opened the hatch, an automatic light turned on. Buzz Aldrin said, "There's that same guy that when you open up the door, he's waiting there for you and he turns the lights on."

Duke said, "How about that? Just like the refrigerator."

Aldrin entered the LM first, taking the TV camera with him. He spun around and pointed it back toward the command module. Duke said, "Hey, that's a great shot. We see you in there. I guess that's Neil and Mike. Better be, anyway."

The broadcast lasted an hour and a half, allowing the astronauts to give a thorough tour of the Lunar Module *Eagle* and some of the equipment they would be using. Mission Control and the outside world watched with great interest.

Late in the broadcast, Armstrong had the camera and was going to great lengths to show the various panels and switches onboard *Eagle*. After Duke commented on the super images they were receiving, Armstrong described his situation as "probably the most unusual position a cameraman has ever had, hanging by his toes from a tunnel and taking the picture upside-down." Duke informed them, "You've got a pretty big audience. It's live in the US. It's going live to Japan, Western Europe, and much of South America. Everybody reports very good color. We appreciate the great show."

Mike Collins remained in the Command Module *Columbia* during the broadcast. Armstrong pointed the camera at him through the tunnel. Duke said, "That's a good view of Mr. Collins down there. We finally see him again."

Collins said, "Hello there, Earthlings."

---

[5] Upon graduating from the US Naval Academy, Charlie Duke failed his Navy commissioning physical exam due to a minor astigmatism in his right eye. "You can go into the Air Force, they will take you," the Navy doctor said. In all the years since, no Air Force or NASA doctor has seen any astigmatism. Charlie Duke, to Southside Community Church, Corpus Christi, Texas, September 6, 2018.

Aldrin said, "It's like old home week, Charlie, to get back in the LM again."

Duke asked if Collins was going in the Lunar Module to look around.

ARMSTRONG: We're willing to let him go, but he hasn't come up with the price of the ticket yet.
DUKE: Rog. I'd advise him to keep his hands off the switches.
COLLINS: If I can get them to keep their hands off my DSKY, it'd be a fair swap.
ALDRIN: That's why I've been eating so much today. I haven't had anything to do. He won't let me touch it anymore.

PAO (WARD): It appears now that we have a view of Earth out the window.

DUKE: 11, Houston. If that's not the Earth, we're in trouble.
ARMSTRONG: That's the Earth, and we have a very good view of it today. There are a few more cloud bands than yesterday when we beamed down to you, but it's a beautiful sight.

Before signing off, Armstrong, an Eagle Scout, expressed a special message. "And Charlie, I'd like to say hello to all of my fellow Scouts and Scouters at Farragut State Park in Idaho. They're having a National Jamboree there this week, and Apollo 11 would like to send them best wishes."

Charlie Duke, also an Eagle Scout, appreciated the gesture: "Thank you, Apollo 11. I'm sure that if they didn't hear that, they'll get the word through the news. Certainly appreciate that." A World Scout Badge was making the journey with Armstrong in his personal preference kit.[6]

After the broadcast ended, Duke thanked them for the wonderful program. "That was one of the greatest shows we've ever seen. We sure appreciate it."

---

[6] "World Scouting Salutes Neil Armstrong," January 1, 2012, https://www.scout.org/node/9739.

PAO (WARD): This is Apollo Control. That television transmission lasted about one hour and thirty-six minutes, according to our first rough calculation. During that period of time, the spacecraft traveled something over 2,000 nautical miles.

Apollo 11's velocity continued to dwindle as it moved farther away from Earth, slowing to a rate of 3,000 feet per second. At 10:00 p.m., it entered the Moon's sphere of influence—"cresting the hill," in the words of FDO David Reed—and began accelerating toward the Moon. The spacecraft was 186,500 nautical miles from Earth and 34,000 nautical miles from the Moon.

A relaxed atmosphere in Mission Control: CAPCOM Charlie Duke takes part in the live television broadcast from Apollo 11 on the third day of the mission. Duke and engineers are looking back at a television monitor set up near the Public Affairs Officer console. The monitor gave a much clearer image than the large screen at the front of Mission Control, visible in the center of the photo. Screenshot from NASA 16mm film graciously provided by Stephen Slater.

17

SO THAT'S WHAT HE WAS DOING WITH THE WORLD BOOK

SATURDAY, JULY 19, 1969

At 8:00 Saturday morning, the Moon passed between the Sun and the Apollo 11 spacecraft, causing an eclipse for the astronauts. After asking for the proper camera exposure setting, Buzz Aldrin described the scene to CAPCOM Bruce McCandless. "We've got the Sun right behind the edge of the Moon now. It's quite an eerie sight. There is a very marked three-dimensional aspect of having the Sun's corona coming from behind the Moon the way it is. I guess what's giving it that three-dimensional effect is the earthshine."

Collins said, "The earthshine coming through the window is so bright you can read a book by it."

Armstrong added, "I'd suggest that along the ecliptic line, we can see corona light out to two lunar diameters from this location. The bright light only extends out about an eighth to a quarter of the lunar radius."

In 1972, Armstrong told astrophysicist Dr. Neil deGrasse Tyson, "My most indelible memory was approaching the Moon and flying through the Moon shadow, so the Moon was eclipsing the Sun. We could see the corona all around the Moon. It was not circular; it was elliptical, which was a big surprise. I didn't understand that." Armstrong continued, "Then we could see the Moon, the dark side of the Moon, of course, illuminated by earthlight. We could see the craters and the valleys and the plains in a blue-gray three-dimensional view that was spectacular. The texture, the image, remarkable but imperceptible to a camera. But from the human eye, it was wonderful."[1]

Later in the morning, it was time to read the day's news. McCandless had a special guest to assist him, Apollo 11 backup LM Pilot Fred Haise. Haise began, "First off, it looks like it's going to be impossible to get away from the fact that you guys are dominating all the news back here on Earth. Even Pravda in Russia is headlining the mission and calls Neil 'the Czar of the Ship.' I think maybe they got the wrong mission."[2]

Later, McCandless said, "In Moscow, space engineer Anatol Koritsky was quoted by TASS as saying that Luna 15 could accomplish everything that had been done by earlier Luna spacecraft. This was taken by the press to mean Luna 15 could investigate the gravitational field, photograph the Moon, and go down to the surface to scoop up a bit for analysis."

Haise finished the morning news with an update on Collins' son, Michael Junior. "Even the kids at camp got in the news when Mike Junior was quoted as replying 'yeah' when somebody asked him if his daddy was going to be in history. Then after a short pause, he asked, 'What is history?'"

After Armstrong thanked them for the "Bruce and Fred Show," Collins said, "You tell Michael Junior, history or no history, he'd better behave himself."

Seven thousand nautical miles from the Moon, Neil Armstrong turned his attention to the extraordinary scene outside the spacecraft.

ARMSTRONG: The view of the Moon we've been having recently is really spectacular. It fills about three-quarters of the hatch window. And we can see the entire circumference, even though part of it is in complete shadow and part of it's in earthshine. It's a view worth the price of the trip.

---

[1] Callum Hoare, "Moon Landing: Neil Armstrong Reveals What Cameras Missed in Unearthed interview," October 2, 2019, https://www.express.co.uk/news/science/1185497/moon-landing-neil-armstrong-apollo-11-cameras-missed-fake-apollo-11-spt.
[2] Pravda was the official newspaper of the Communist Party of the Soviet Union.

McCANDLESS: Well, there are a lot of us down here that
would be willing to come along.
COLLINS: I hope you get your turn, and soon.
ARMSTRONG: One of these days, we'll be able to bring the
whole MOCR along, I hope. That will save a lot of antenna
switching.

After noon, Apollo 11 slipped around the western edge of the Moon
losing radio contact with Houston. Soon afterward, the SPS engine
burned for six minutes slowing the spacecraft and enabling it to be
captured by lunar gravity. The engine shut down right on time, and the
preliminary numbers looked good. Collins said, "I take back any bad
things I ever said about MIT, which I never have."

After the burn, the crew had a few moments to look out the window
at the far side of the Moon, which was illuminated by the Sun. They
weighed in on a lingering question from Apollo 8 and 10 involving the
actual color of the Moon.

Aldrin said, "Well, I have to vote with the Apollo 10 crew. That
thing is brown." Collins agreed.

Armstrong said, "Looks tan to me."

Aldrin and Collins mentioned when the Sun was near the horizon,
the surface looked gray but turned brown as the Sun moved overhead.

The first order of business was asking the computer for results of
their Lunar Orbit Insertion burn. The target was an elliptical orbit of
169.2 by 61.0 nautical miles. Aldrin read the actual outcome. "Look at
that! 169.6 by 60.9. We only missed by a couple of tenths of a mile."

The crew resembled tourists as they peered out the different
windows in the Command Module. The rarely-excited Armstrong said,
"What a spectacular view!"

Collins said to Armstrong, "Yes, there's a moose down here you
just wouldn't believe. There's the biggest one yet. It is enormous! It's
so big I can't even get it in the window. You want to look at that? That's
the biggest one you've ever seen in your life. Neil, look at this central
mountain peak. Oh, boy, you could spend a lifetime just geologizing that
one crater alone, you know that?"

"You could."

"That's not how I'd like to spend my lifetime, but picture that.
Beautiful!"

The crew gained radio communication with Houston after seeing earthrise. They reported on the successful burn, then described the terrain leading to their landing site.

ARMSTRONG: Apollo 11 is getting its first view of the landing approach. At this time, we are going over the Taruntius Crater, and the pictures and maps brought back by Apollos 8 and 10 have given us a very good preview of what to look at here. It looks very much like the pictures, but like the difference between watching a real football game and one on TV, there's no substitute for actually being here.
McCANDLESS: Roger. We concur, and we certainly wish we could see it first hand also.
ARMSTRONG: We're going over the Messier series of craters at this time, looking vertically down on them, and in Messier A, we can see good-sized blocks in the bottom of the crater. I don't know what our altitude is now, but in any case, those are pretty good-size blocks.
ARMSTRONG: There's Secchi in sight. We're over Mount Marilyn at the present time, and the powered descent ignition point.
McCANDLESS: Jim is smiling.[3]

Then Armstrong commented on the features they would fly over during the powered descent. "Currently going over Maskelyne and Boot Hill, Duke Island, Sidewinder. Looking at Maskelyne W—that's the yaw-around checkpoint—and just coming into the terminator."

By "yaw-around checkpoint," Armstrong referred to the astronauts spending the first few minutes of the descent checking lunar landmarks before Eagle rotated 180 degrees allowing the landing radar to point to the Moon's surface.

Then Armstrong weighed in on the Apollo 8 vs. Apollo 10 debate on whether the Moon's color is gray or brown. "At the terminator, it's ashen gray. As you get further away from the terminator, it gets to be a lighter gray. And as you get closer to the subsolar point, you can

---

[3] Jim Lovell named the distinctive feature in honor of his wife during Apollo 8. Lovell, the backup Commander for Apollo 11, sat beside McCandless during this conversation (*see* NASA 16mm video). He sat beside CAPCOM Charlie Duke during the landing.

definitely see browns and tans on the ground, according to the Apollo 11 observation anyway."

A disappointed McCandless replied, "We're recording your comments for posterity." His mic picked up someone sarcastically calling them "compromisers." Some in Mission Control wanted them to take one side of the argument or the other, not merge them.

The comment did not sneak past Mike Collins, who asked, "Did somebody in the background accuse us of being compromisers?"

As usual, Armstrong's description was definitive. The next day during the final moments of the landing, the world would see what the NASA team knew well—he was anything but a compromiser.

On their next orbit, the crew began another television broadcast as they emerged from the far side of the Moon. The Sun was high overhead; there were no shadows, and the surface looked brown and washed out.

McCandless said, "We're getting a beautiful picture in down here now, 11. The color's coming in quite clearly, and we can see the horizon and the relative blackness of space, and without getting into the question of grays and browns, it looks, at least on our monitor, sort of a brownish-gray."

They approached Mare Fecunditatis, the Sea of Fertility. On its eastern shore is the magnificent crater Langrenus, 130 km in diameter, with terraced walls and mountain peaks on the crater floor. Located near the eastern limb of the Moon, its circular shape appears oblong as viewed from Earth.

ALDRIN: We're moving the camera over to the right window now to give you Langrenus with its several central peaks.
McCANDLESS: We're getting a beautiful picture of Langrenus now with its rather conspicuous central peak.
COLLINS: The Sea of Fertility doesn't look very fertile to me. I don't know who named it.
ARMSTRONG: Well, it may have been named by a gentleman whom this crater was named after, Langrenus. Langrenus was a cartographer to the King of Spain and made one of the early reasonably accurate maps of the Moon.
McCANDLESS: Roger. That's very interesting.
COLLINS: I'll have to admit it sounds better for our purposes than the Sea of Crises.
McCANDLESS (LAUGHING): Amen to that.

Janet Armstrong watched intently in her Houston home. While her husband gave the history lesson, she thought, "So *that's* what he was doing with the World Book Encyclopedia in his study!"[4]

Mike Collins commented again on the color of the Moon. "At these low Sun angles, there's no trace of brown. It's now returned to a very gray appearance and, like the Apollo 8 crew said, it has a look of Plaster of Paris at this Sun angle, which is completely lacking at higher Sun angles."

The camera zoomed in on Mount Marilyn, where the Powered Descent Initiation (PDI) would occur. Then they flew over the steep ridges which impressed Tom Stafford and Gene Cernan from a low altitude on Apollo 10. Armstrong mentioned other features such as Boot Hill, which they would see three minutes and fifteen seconds into the descent. He said, "The largest of the craters near the center of the picture right now is Maskelyne W. This is a position check during descent at about three minutes and thirty-nine seconds, and it's our downrange position check and cross-range position check prior to yawing over face-up to acquire the landing radar. Past this point, we would be unable to see the surface below us until getting very near the landing area."

He mentioned the rille which the Apollo 10 crew referred to as Sidewinder. Collins said, "That's a good name too, Sidewinder and Diamondback. It looks like a couple of snakes down there in a lakebed."

They approached darkness. Aldrin said, "It's a very fantastic view to see the terminator as you look along the edge of it. I think you'll agree that some of these craters that you're seeing in the picture now are really accentuated by the lengthening of the shadows as they approach the terminator."

Buzz Aldrin drew the telecast to a close. "And as the Moon sinks slowly in the west, Apollo 11 bids good day to you."

McCandless said, "We sort of thought it was the Sun setting in the east."

Collins said, "Well, it depends on your point of view."

The crew returned to business as they prepared for landing day. They performed a short 17-second burn to circularize their orbit. Armstrong and Aldrin re-entered the LM for some housekeeping and to

---

[4] Armstrong, Collins, Aldrin, *First on the Moon*, 238.

check its communications with Houston. At the same time, Collins performed some tracking exercises on the frontside to give them more data about their orbits. At 11:00 p.m. Houston time, the crew signed off and tried to rest before the big day.

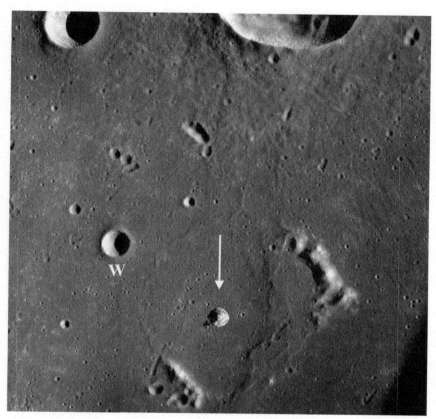

Apollo 10 Command Module Charlie Brown pictured between key landmarks leading to the Apollo 11 landing site: Boot Hill (right), Duke Island (lower left), and Maskelyne W. The large crater at the top right of frame (North) is Maskelyne.
Photo credit: NASA AS10-29-4170,
annotations by author.

Mount Marilyn, point of ignition for the Apollo 11 descent. Weatherford Crater, named by Tom Stafford during Apollo 10, is left-center near the top. Image from the Lunar Reconnaissance Orbiter Camera (LROC). Photo credit: NASA/GSFC/Arizona State University.

Approach to the Apollo 11 landing site, looking west. Maskelyne W provided a key position check early in the descent. Sabine E was later renamed "Armstrong," Sabine D was renamed "Collins," and Sabine B (in shadow) was renamed "Aldrin" by the International Astronomical Union. Photo credit: NASA AS11-37-5437, annotations by author.

18

A GUY NAMED BOB WHITE CAME UP WITH THIS WEIRD THING, AND WE THOUGHT HE WAS CRAZY

SUNDAY MORNING, JULY 20, 1969, 6:05 a.m.
9 HOURS BEFORE POWERED DESCENT INITIATION (PDI)

Ron Evans, CAPCOM for Glynn Lunney's Black Team, gave a wake-up call to Mike Collins. The spacecraft would complete its ninth frontside pass and lose communication with Houston in two minutes.

> EVANS: Apollo 11, Apollo 11. Good morning from the Black Team.
> COLLINS: Good morning, Houston. You guys wake up early.
> EVANS (LAUGHING): Looks like you were really sawing them away.
> COLLINS: How are all the CSM systems looking?
> EVANS: Looks like the Command Module is in good shape. Black Team has been watching it real closely for you.
> COLLINS: We sure appreciate that. Because I sure haven't.

Thus began one of the most intense days of their lives. During the next hour, Gene Kranz's White Team members drifted into the Mission

Operation Control Room on the third floor of Building 30 before their
shift change. Some, like Kranz, spent the previous night at home with
family and went to church before reporting for duty. Others, like Steve
Bales, spent the night in the flight controller bunkroom located on the
second-floor walkway between the office section of Building 30 and the
Mission Control area.[1]

Inside the spacecraft, the crew ate a large breakfast on their tenth
orbit, knowing it would be a while before they could eat again.[2] Ron
Evans read them the morning news during their meal:

> EVANS: Church services around the world today are
> mentioning Apollo 11 in their prayers. President Nixon's
> worship service at the White House is also dedicated to the
> mission. And our fellow astronaut, Frank Borman, is still in
> there pitching and will read the passage from Genesis, which
> was read on Apollo 8 last Christmas.[3] The Cabinet and
> members of Congress, with emphasis on the Senate and
> House Space Committees, have been invited along with a
> number of other guests. Buzz, your son, Andy, got a tour of
> the Manned Spacecraft Center yesterday. Your Uncle Bob
> Moon accompanied him on the visit, which included the LRL
> [Lunar Receiving Laboratory].[4]
> ALDRIN: Thank you.
> EVANS: You residents of the spacecraft *Columbia* may be
> interested in knowing that today is Independence Day in the
> country of Colombia…. Even though research has certainly
> paid off in the space program, research doesn't always pay
> off, it seems. The Woodstream Corporation, parent company
> of the Animal Trap Company of America, which has made
> more than a billion wooden-spring mousetraps, reports that it
> built a better mousetrap, but the world didn't beat a path to
> its door. As a matter of fact, the company had to go back to
> the old-fashioned kind. They said, "We should have spent
> more time researching housewives and less time researching
> mice."

---

[1] Johnson Space Center visitors touring the renovated Mission Control will see the
walkway with an irregular edifice just past the entrance to Building 30.
[2] NASA, "Apollo 11 Technical Crew Debriefing," 8-11.
[3] Kurson, *Rocket Men*, 281. "Stay in there and pitch," meaning "Don't Give Up," was
the motto of Frank Borman's father during their family's bleak days of the Depression.
[4] The crew would spend three weeks quarantined in the LRL after the mission.

From then on, it was all business. After the spacecraft went behind the Moon, Kranz's team began its shift in the Control Center. Kranz saw Bill Tindall in the viewing area and promptly waved Tindall down to sit beside him. Bales, the Guidance Officer, took his seat in the front row. To his right was Gran Paules, another Guidance Officer who would be assisting him during the landing. To Bales' left was Jay Greene, the Flight Dynamics Officer (FDO) for the descent. Bales and Greene would work closely throughout the day. Between Bales and Greene sat their branch chief Jerry Bostick ready to help with the Mission Rules and his own copy of the computer alarm codes which Garman had compiled.[5]

In the Flight Dynamics Staff Support Room, commonly called the "Back Room," were specialists helping the flight controllers. Jack Garman sat on the front row at the far-right console, supporting Bales as he did in the simulations. Sitting beside Garman was Ken Baker and his assistant Glen Shook, experts from TRW Systems Group, which developed the Abort Guidance System.[6] Also with Garman was Russ Larson, a computer expert from MIT.

Over 1500 miles away in Cambridge, Massachusetts, Larson's co-workers gathered in a second-floor MIT classroom listening to the air-to-ground transmissions on a squawk box.[7] Included in the group were Allan Klumpp and Don Eyles, who had written much of the software controlling the Lunar Module's descent.

While the White Team settled into their positions, Armstrong and Aldrin migrated into *Eagle* on the Moon's far side and began the complicated LM activation and checkout process. By 8:30 a.m. Houston time, they had successfully powered up the Lunar Module. When they acquired communications for the 11th orbit, Mission Control specialists began to evaluate Eagle's condition. The LM components were designed to survive the violent ordeal of a Saturn V launch, undergoing a battery of pre-flight vibration tests and thermal tests, but each system had to be carefully assessed after power up.[8]

---

[5] Jerry Bostick, interview by author, August 25, 2020.
[6] Steve Bales, email message to author, March 2, 2020.
[7] Don Eyles, "Tales from the Lunar Module Guidance Computer," A paper presented to the 27th annual Guidance and Control Conference of the American Astronautical Society (AAS), in Breckenridge, Colorado, February 6, 2004, https://www.doneyles.com/LM/Tales.html.
[8] Steve Bales, interview by author, October 9, 2019.

Once the guidance computer on *Eagle* was up and running, Steve Bales and Jay Greene turned their attention to the major challenges: detecting the health of the LM guidance platform and giving the LM computer the most accurate information for the powered descent. Since the onboard guidance computer would fly the LM on autopilot for most of the descent, both tasks were critical.

To check the health of the guidance system, Bales needed to compare two alignments separated by a period of time. It would indicate whether the platform remained stable or drifted more than expected. They had to perform the first alignment while the two vehicles were docked. The second would be done hours later after the LM undocked before the landing attempt.

The opportunity for the first alignment came sooner than expected. The crew was half an hour ahead of their timeline on the 11th orbit. Aldrin headed back into the CM to don his suit. About 25 minutes before they lost communication with Houston, Armstrong decided to perform the docked alignment even though it was scheduled for the next revolution on the flight plan. Squeezing it in early provided an extra hour of data and a more conclusive result. "This gave us better drift checks, which was a help in analyzing the LM platform," said Armstrong.[9]

DOCKED ALIGNMENT

Getting an accurate Lunar Module alignment while docked to the Command Module was challenging. *Eagle* had a precise Alignment Optical Telescope (AOT), but it was blocked by *Columbia* when joined nose to nose. Apollo 10 worked around this by using mechanical equations to translate the known CM platform alignment to the LM for a coarse (rough) alignment. Slippage and other complications during Apollo 10 made mission planners realize a more precise docked alignment was necessary for Apollo 11. Once again, NASA turned to MIT for a solution.

In a May 15, 1969 memo, Tindall conveyed confidence in a new procedure developed by Bob White of MIT:

---

[9] NASA, "Apollo 11 Technical Crew Debriefing," 8-14.

The technique requires two spacecraft attitude maneuvers while in the docked configuration with the LM and CM crew simultaneously keying out [gimbal] angles before and after each of these attitude changes. All of this must be done after the LM IMU [platform] has been coarsely aligned as in the current flight plan. With this data, the flight controllers can compute the LM IMU orientation and torquing angles required. This technique is expected to be as good as an AOT alignment.[10]

Tindall was not quite as confident when he first heard about the plan. Steve Bales put it bluntly:

I give MIT a lot of credit. A guy named Bob White came up with this weird thing, and we thought he was crazy. Tindall thought he was too. White said if you maneuver these docked vehicles a certain way and take these different readings on their platforms, you can get a very, very accurate docked alignment. We finally told White to give us the equations and prove to Bill that this is going to work, and darned if it didn't.[11]

It was quite a complicated process. The first step was calculating a preferred landing alignment so the cockpit roll, pitch, and yaw indicators would each read zero when the LM touched down on the lunar surface. The main display in the cockpit was the black and white Flight Director Attitude Index (FD/AI, commonly called the 8-ball), similar to the Attitude Indicator or false horizon used by pilots in an airplane. The bottom half of the 8-ball was black, the top half white, with distinctive vertical lines and dashes on each half. It was most helpful when the spacecraft was oriented near its desired alignment because the astronauts could easily see how their current attitude compared to what it should be. Surrounding the 8-ball were indicators for the roll rate (above the 8-ball), pitch (on the right), and yaw (underneath).

The Lunar Module sitting on the Moon provided a reference to a known position and was called a "Reference to Stable Member Matrix,"

---

[10] Bill Tindall, "Some 'Improvements' in Descent Preparation Procedures," Tindallgram, May 15, 1969,
http://www.collectspace.com/resources/tindallgrams/tindallgrams01.pdf.
[11] Steve Bales, interview by author, November 6, 2019.

or REFSMMAT.[12] There were several REFSMMATS used on a lunar mission, including an alignment in reference to the launch pad at the Cape, one for the trip out to the Moon, one for the burns in lunar orbit, one for landing, one for lunar launch, one for traveling back to Earth, and one for entry. By having a REFSMMAT for each major phase of the mission, the 8-ball orientation could be kept in the neighborhood of where it was supposed to be for critical maneuvers, enabling the astronauts to assess how close they were to their desired attitude quickly.

At the beginning of his shift, Steve Bales had to calculate the preferred REFSMMAT for landing on the Moon. To do this, Jay Greene calculated the projected landing time from the current trajectory. Bales added the landing coordinates and gave the information to the computer operator in the Real-Time Computer Complex (RTCC) on the first floor of the building. The computer used a program called Lunar Surface Align Display (LSAD) to calculate the preferred REFSMMAT for the landing. Bales recalls, "I had written the requirements for the LSAD based on knowing what would be required both in orbit and landing on the surface, much of which was learned via the Tindall meetings. Then others wrote the equations and IBM programmers coded them for use on the MOCR computers."[13]

The LSAD then passed the preferred REFSMMAT to a command program that built upload links. Gran Paules sent one to the Command Module and a corresponding one to the Lunar Module. The computers on the CM and LM could store two REFSMMATs, one they were currently aligned to, and also the preferred REFSMMAT for the next desired maneuver. The computer could smoothly adjust the spacecraft's orientation from one to the other.[14] For the purpose of aligning *Eagle*, Mike Collins switched *Columbia* to the preferred REFSMMAT which Paules uploaded.

Once the CM was aligned to the preferred REFSMMAT, the same alignment was transferred to the LM through a coarse alignment procedure. Collins held the attitude of the docked spacecraft steady as he and Armstrong punched entries into their keyboards, calling up the

---

[12] REFSMMAT was part of the "Twas the Night Before Christmas" poem read to the Apollo 8 crew by Harrison Schmitt. *See* Chapter 10, page 124.

[13] Steve Bales, email message to author, November 23, 2019.

[14] Steve Bales, interview by author, November 6, 2019.

current gimbal angles on their displays. On Armstrong's "Mark," Collins read his angles to the Commander, who torqued (adjusted) *Eagle's* guidance platform to the corresponding angles.

Next came the fine alignment, although not as accurate as Bob White's equation would be. Armstrong counted down, "Three, two, one," and on his "Mark," he and Collins called up their current gimbal angles on the display. Armstrong asked if Houston received those angles. CAPCOM Charlie Duke replied, "*Eagle*, Houston. We have the angles. I'll read them back. For the Command Module: 11154, 20792, 00230. For the LM: 18995, 02852, 35863. Over."

Those numbers assume a decimal after the third number, so the CM roll was 111.54 degrees, pitch 207.92 degrees, and yaw 2.30 degrees. And for the LM, roll 189.95 degrees, pitch 28.52 degrees, and yaw 358.63 degrees. Bales notes, "Theoretically, the pitch readings would be 180 degrees apart, but some bending must have occurred. We took those angles and entered them into another program, the Lunar Orbital Alignment Program, which computed the torquing angles. I filled out a Pre-Advisory Data form and took it to Charlie, who read it up to Neil."[15]

The alignment was good, but not good enough. Now Bob White's procedures came into play. Collins maneuvered the docked spacecraft to a specific attitude. Armstrong and Collins called up the angles on the display as they did earlier, and the ground recorded them. Collins maneuvered the spacecraft to a different orientation, and they repeated the same process. Again, the angles were recorded. Steve Bales relayed those values to the computer operator who entered them into the Lunar Orbital Alignment Program, which computed the torquing angles. About six minutes before Loss of Signal, Charlie Duke told the crew the angles looked good and they could torque *Eagle's* platform, producing what Bales hoped would be a precise alignment for the Lunar Module. "Again, I only wrote the requirements; Bob White supplied the equations and IBM programmed them."[16] Bales had to wait several hours before he learned if the process actually worked.

While Bales focused on the alignment, Greene evaluated *Eagle's* orbit. They needed to send all the trajectory data to the onboard

---

[15] Steve Bales, email message to author, November 23, 2019.
[16] Steve Bales, email message to author, November 23, 2019.

computer at 1:00 p.m. as it came around the edge of the Moon on its 13th orbit. The landing attempt would begin on the following revolution, but the final minutes before the descent would already be too saturated with other procedures to send up the data to the crew.

That gave the 11th and 12th frontside passes to get it right. The trajectory data needed to be as precise as possible since what they uplinked at 1:00 p.m. had to accurately project the LM's position, orientation, and velocity at 3:00 p.m. when the powered descent would begin. The landing radar could see the lunar surface and provide accurate real-time information three and a half minutes into the descent. Until then, the computer would rely on the data uplinked two hours earlier, with any deviations causing problems during the descent.

The Moon did not make it easy to predict where the LM would be on future revolutions. Its irregular concentrations of mass below the lunar surface wreaked havoc on the spacecraft's orbits.

Jay Greene gathered trajectory data for the 70 minutes the spacecraft was on the near side of its 11th orbit. He fed it to Bales, and while *Eagle* and *Columbia* were on the far side of the Moon, the computers on the first floor of their building ground out the predicted orbit. When they acquired communications with Houston, Gran Paules uplinked the trajectory information to the LM onboard computer telling it where it was, where it was going, and its velocity. This information was contained in a series of numbers called the "state vector."

They repeated the process during the 12th revolution, fine-tuning their predictions as they added more and more information to their storehouse of knowledge. They knew they would never get the elusive orbit exactly right, but they had to get it as close as possible. They were confident as the spacecraft went behind the Moon for the 12th time.

UNDOCKING, 12:45 p.m.

The computers made their calculations again while the spacecraft was on the far side of the Moon. *Columbia* undocked from *Eagle* before acquiring communication for the 13th orbit. Collins filmed the event with a Data Acquisition (movie) camera and his Hasselblad. As *Columbia* and *Eagle* appeared for the 13th time at 1:00 p.m., Armstrong and Aldrin set the onboard computer to Program 00 (P00), giving control of the computer to Houston. Gran Paules uplinked the state vector information directly to the LM computer. Additionally, Duke read the Pre-Advisory Data (PAD) for the Powered Descent Initiation (PDI) burn which would begin the landing attempt on the next revolution, plus PADs for abort maneuvers if the LM had to return to the Command Module.

The two spacecraft flew in close formation 60 miles above the lunar surface. Duke asked, "How does it look, Neil?"

"The *Eagle* has wings."

Collins inspected *Eagle*, making sure the landing gear was extended and locked. He said, "I think you've got a fine-looking flying machine there, *Eagle*, despite the fact you're upside-down."

Armstrong said, "Somebody's upside-down."

"Okay, *Eagle*. One minute until ignition. You guys take care."

"See you later."

"I hope so," thought Collins.[17]

Collins fired his thrusters, gradually dropping *Columbia* below *Eagle* at 2.5 feet per second, giving them plenty of separation for the upcoming maneuvers. Aldrin photographed *Columbia* as they flew over the landing site for the last time before the attempt with familiar landmarks in the background. Armstrong told Collins, "You're going right down US 1, Mike."[18]

It was day six of the lunar month. The Sun was rising on the southwestern part of the Sea of Tranquility, and a little less than half of the Moon's near side was illuminated. Soon after they passed the landing site, the two spacecraft crossed the terminator into darkness.

Now he could see stars, and it was time for Armstrong to perform the long-awaited fine alignment of the LM using the Alignment Optical Telescope (AOT). This would determine how much the LM platform had drifted in the four hours since the complicated docked alignment. Armstrong used *Eagle's* AOT to mark two stars, Acrux and Antares, carefully. The flight controller displays in Houston revealed he was taking five marks on each star, more marks than the procedure called for, but it increased the accuracy of his measurements. Steve Bales

---

[17] Collins, *Carrying the Fire*, 398.

[18] US Highway 1 is a main north-south highway in the eastern United States. It was a nickname the astronauts gave Rima Hypatia, a long rille (groove, channel) that served as a landmark for the Apollo 11 landing site.

anxiously awaited the results, fearing an unstable platform would be his worst nightmare of the day.

The process was taking longer than usual. Flight controllers observed an alarm occurring inside the cockpit, indicating Armstrong was taking more marks than necessary. Gene Kranz knew the significance of this alignment and asked Bales how much longer it would take. Bales let him know the Commander was being meticulous.

The silence was deafening as they waited for Armstrong to complete the process. Fantastic news jumped out of Bales' display. The gimbal angles were right on target; the platform alignment remained extremely precise. He excitedly shared the results with Kranz, who chuckled at his exuberance. Then Bales shouted, "GREAT Torquing Angles!" With confirmation *Eagle's* guidance platform was healthy, Bales' biggest hurdle of the day was behind him. Or so he thought.

Charlie Duke was listening to all of this. Knowing he could not communicate anything to the crew without the Flight Director's approval, Duke dutifully asked if he could inform Armstrong and Aldrin to "torque" the gimbal angles, telling the computer to accept the new angles as correct.

"Do we accept them?" he asked Kranz and Bales.

An excited Bales replied, "Affirmative!" Kranz chuckled again, and Duke passed it along, "Rrroger, *Eagle*, you can torque it, over."

Bales was not the only one watching with great interest while Armstrong performed the alignment. Flight Activities Officer (FAO) Spencer Gardner and his team developed the flight crew's step-by-step procedures, checklists, and schedules for the entire flight, including Bob White's complicated method. When Armstrong confirmed the accuracy of the docked alignment, Gardner spiked the football. He calmly spoke to his support room, "I told you guys the docked align would be good."

Bales points to this event as an illustration of the teamwork involved in mission success:

> While Tindall's group developed the concept, Bob White came up with the theory and equations. I wrote the requirements for the Control Center software and was responsible for the overall activity since the guidance system depended on it. Spencer and his team were in charge of developing the actual crew procedures to accomplish it, that

is, when to call up the correct DSKY readings, count down together, record the values, etc., all the many tasks and on a timeline that made this work. It was a real team effort: Flight Techniques, MIT, Guidance Officer, FAO, and support team... one of many things that made the descent work.[19]

Two minutes later, Spencer Gardner called Bales on the Guidance loop. "Maybe nobody will knock that docked alignment anymore."

Bales replied, "Stand by till the next mission. I'm sure they will again."

Five decades later, Bales reflected on the "Great Torquing Angles" moment. "The two points I remember best are this one and the speech Gene gave just prior to Acquisition of Signal before the descent. I can remember the descent, of course, but not quite as clearly. Those 12 minutes were more like a slow-motion wreck."[20]

Buzz Aldrin in Eagle. Photo credit: NASA AS11-36-5390.

---

[19] Steve Bales, email message to author, December 10, 2019.
[20] Steve Bales, email message to author, November 8, 2019.

Command Module Columbia heading west (left) over the landing site.
Sabine B (Aldrin) and Sabine D (Collins) are visible. Sabine E (Armstrong)
is out of view to the right, and the primary crater Sabine is out of view to the
left. Satellite craters are denoted A, B, C, D, E, etc. as they progress in
distance from their primary crater. Lunar landmark 130 is the
small crater on the rim of Sabine D at the 8:30 position.
Photo credit: NASA AS-11-37-5447, annotations by author.

19

A SLOW-MOTION WRECK

THE CALM BEFORE THE STORM

The two spacecraft disappeared behind the Moon at 2:00 p.m. Eight minutes later, half a revolution from where the powered descent would begin, *Eagle* fired the descent engine for 30 seconds in a Descent Orbit Insertion (DOI) burn. This changed its orbit from circular to elliptical, with the low point being an altitude of 50,000 feet. The Lunar Module engine would fire from there, beginning the landing phase.

Mission planners agreed on that altitude years earlier. NASA wanted to begin the descent as low as possible to decrease the amount of propellant needed but were not comfortable beginning lower than 50,000 feet. At that time, they were not sure of the exact radius of the Moon or the height of the mountains they would fly over. They added a little margin for safety and the result was 50,000 feet. Steve Bales explains, "That decision was made in the early 1960s before I got into the business of flight control. They must have had big fights about that. 'Well, should we start at 50,000 feet? Should we start at 40,000 feet?' That is the kind of trade-off we had all the time."[1]

Collins was alone in *Columbia*. Around his neck was a "solo book" containing a map of the landing area, the flight plan for the next day, and procedures for 18 different rendezvous procedures to cover any conceivable situation.

---

[1] Steve Bales, interview by author, October 9, 2019.

The controllers took a break while Apollo 11 was behind the Moon. There was a stark contrast between Houston's young flight controllers averaging 26 years of age, and the veteran pilots in *Eagle*. The controllers were tense, flocking to the restroom during their brief break. Meanwhile, Armstrong and Aldrin were calmly chatting as if they were waiting for another pre-flight simulation. When Armstrong asked for a lunar map, Aldrin replied, "Trade you that for a piece of gum." When interference between the Lunar Module and Command Module radios made an eerie sound in their headsets as it did during Apollo 10, Aldrin laughed and said, "You hear that too, huh?"

Armstrong said, "Sounds like wind whipping around the trees."[2]

Armstrong believed they had a 50 percent chance of a successful landing.[3] He was a lot more optimistic than the flight controllers. Jay Greene anticipated an abort:

Nobody, nobody on the team believed that we'd make it down the first time; I don't think anybody did. That became particularly bothersome for a Flight Dynamics guy because the first part of the job was monitoring the descent. The other part of the job was if we abort, computing all the rendezvous maneuvers to get the LM back with the Command Module. So, it was intense.[4]

Kranz could see the tension growing as the controllers returned from their break. He ordered the Control Center doors to be locked. 26-year-old Steve Bales wondered if it was to keep people out or make sure the flight controllers did not try to escape.[5] Then Kranz instructed them to go to a private communications loop used for debriefings. Only the flight controllers experienced what came next. Bales called it "a speech

[2] *See* Collins, *Carrying the Fire*, 413.
[3] Andrew Chaikin, *Voices from the Moon*, with Victoria Kohl, (New York: The Penguin Group, 2009), 47.
[4] Jay Greene, interview by Sandra Johnson for NASA JSC Oral History, November 10, 2004, 19.
[5] A/V Geeks, "Flying the Apollo Missions, Part 1," Johnson Space Center, December 26, 2013, Video, 1:01:27, https://www.youtube.com/watch?v=kFxlw1Vsy18.

that General Patton would have liked to have given."[6] Kranz recalls his comments on the epic occasion:

> Okay, all flight controllers, listen up. Today is our day, and the hopes and dreams of the entire world are with us. This is our time and our place, and we will remember this day and what we do here always. In the next hour, we will do something that has never been done before. We will land an American on the Moon. The risks are high—that is the nature of our work. We worked long hours and had some tough times, but we have mastered our work. Now we are going to make this work pay off. You are a great team, one that I feel privileged to lead. Whatever happens, I will stand behind every call that you will make. Good luck, and God bless us today.[7]

It was precisely the message they needed to hear. The youthful controllers would remember his speech for the rest of their lives.

22 MINUTES BEFORE POWERED DESCENT

> *PUBLIC AFFAIRS OFFICER (DOUG WARD): This is Apollo Control at 102 hours, 12 minutes into the flight of Apollo 11. We're now two minutes and fifty-three seconds from reacquiring the spacecraft. Twenty-one minutes and twenty-three seconds from the beginning of the powered descent to the lunar surface. It's grown quite quiet here in Mission Control. A few moments ago, Flight Director Gene Kranz requested that everyone sit down, get prepared for events that are coming, and closed with a remark, "Good luck to all of you."*

Doug Ward was well acquainted with the situation. John McLeaish, branch chief of NASA's Public Affairs Office, assigned Ward to the role of mission commentator for the descent phase months earlier. Ward made it a point to be in Mission Control for every integrated sim so he

---

[6] A/V Geeks, "Flying the Apollo Missions, Part 1," Johnson Space Center, December 26, 2013.
[7] Kranz, *Failure Is Not an Option*, 283–284.

could describe the action as accurately as possible.[8] He especially liked working with Kranz:

> Gene was the quintessential engineer. With Kranz, you knew exactly what was going on because he would work a problem literally to death. Some of the other Flight Directors would get involved in what I call "the verbal shorthand," the technical discussion between themselves and the flight controllers. And for a guy like me, a public affairs commentator, it was sometimes difficult to follow what was going on because they were using their own technical shorthand so much. And so much of the interchange was understood. It was not articulated because they each knew what the implications of each particular piece of data would be, and they would not discuss it. They just knew intuitively. So, you often did not get that clear elucidation of what an issue was. It tended to be the other way with Kranz because he worked it so thoroughly over and over. I always felt with Gene, I had a better understanding of what was going on than I did with almost any other Flight Director.[9]

At 2:47 p.m., CAPCOM Charlie Duke heard from Mike Collins as *Columbia* came around the east limb of the Moon. *Eagle* was ahead of *Columbia* due to its lower orbit but was so close to the surface that Collins came into radio contact first.[10] The drama was about to begin.

Duke asked how the Descent Orbit Insertion burn went.

COLLINS: Listen, babe. Everything's going just swimmingly. Beautiful.
DUKE: Great. We're standing by for *Eagle*.
COLLINS: Okay. He's coming along.

Soon *Eagle* came into range, and Aldrin reported the DOI burn went as planned. The LM was currently 18 nautical miles (109,000 feet)

---

[8] Doug Ward, interview by author, March 16, 2019.
[9] Doug Ward, interview by author, March 16, 2019.
[10] *Eagle* moved about 120 miles ahead of *Columbia*. Conversely, when *Eagle* fired the descent engine and slowed, *Columbia* "would whiz by overhead," as Mike Collins described, moving about 200 miles ahead when *Eagle* landed. Collins, *Carrying the Fire*, 398.

above the lunar surface, dropping at a rate that would cut their altitude in half by the time powered descent would begin.

Gene Kranz had two opportunities to attempt the descent. Kranz could wave it off if things did not look favorable with the first opportunity. The LM would remain in a safe orbit, and they could try again on the next revolution. If he waved off the second attempt, the crew would have to come home empty-handed. "The Moon was rotating underneath us and the landing site was shifting out of plane. We could only accommodate so much plane change," explained David Reed.[11] The exact landing coordinates were such an integral part of the Lunar Module's computer program that it would not be possible to land at a random spot on the Moon.

After Aldrin gave the DOI burn report, Mission Control lost all data from *Eagle*. Duke used Collins as a relay to communicate with his friends in the landing module, asking them to try the high-gain antenna. A minute later, communications were restored between *Eagle* and Houston.

12 MINUTES BEFORE PDI, POWERED DESCENT INITIATION

> *PAO (WARD): Everything still looking very good at this point. We presently show the LM at an altitude of 12.9 nautical miles and descending.*

The astronauts were flying west, feet first, and looking down toward the Moon. Armstrong wanted to see the lunar surface for the first few minutes of the descent so he could note the time they passed over landmarks and get an idea of where they would land in relation to their aiming point. The disadvantage was the landing radar, located on the opposite side of the LM from the main windows, pointed away from the lunar surface until *Eagle* rotated three and a half minutes into the descent. Prior to that maneuver, the guidance computer would rely on the data Bales and Greene uplinked two hours earlier, predicting the details of *Eagle's* orbit.

On Apollo 12 and all future lunar missions, Commanders elected to begin their descents with the radar pointing toward the lunar surface,

---

[11] David Reed, email message to author, January 10, 2020.

much to the relief of Bill Tindall and his Mission Techniques team. Armstrong and Aldrin were in a unique situation, blazing a trail with many unknowns regarding how the LM guidance system would respond. Armstrong wanted confirmation from visual cues. Not everyone agreed with the decision.

In the planning stages, the Mission Techniques group argued for the descent to start with the landing radar pointing toward the surface so it could accurately read the altitude and velocity as soon as possible. Steve Bales recalls:

> We were on a campaign. "We want the radar. We want the radar!" Neil comes to one of our meetings and says, "I want to see the ground because I can tell you if we're on the right track or not." Nobody in flight techniques other than Neil and maybe Buzz thought he could do it. I say "nobody," but at least the guys who were going to monitor it didn't. Jay and I didn't. Tindall didn't. Bill even noted, "I don't think he can." But Neil won the day on that.[12]

A few minutes before powered descent, Armstrong used what he called "Barnyard Math" to calculate their altitude by measuring the length of time it took for landmarks to pass between marks on his window and comparing it to a chart prepared in advance.[13] These backup checks confirmed the LM was at the correct altitude.

In Mission Control, Charlie Duke noticed an alarm from the LM and asked Steve Bales about the cause. As Bales looked into it, a blast of static came through the air-to-ground communications loop, making it difficult for them to hear each other. Gene Kranz was anxious:

> We don't have to look for problems because they come right at us, like flies drawn to a picnic lunch. Voice communications

---

[12] Steve Bales, interview by author, November 6, 2019. Jerry Bostick remembers Armstrong fondly. "Neil was something else. He didn't talk a lot, but when he did, you should pay attention to what he had to say. Even if you didn't understand it at the time, we would go away saying, 'Hmm, let's think some more about this.' He was a real engineer." Jerry Bostick, interview by author, August 25, 2020.
[13] Neil Armstrong, interview by Katy Vine, "Walking on the Moon," TexasMonthly, July 2009, https://www.texasmonthly.com/articles/walking-on-the-moon/. Katy Vine interviewed many key Apollo veterans, producing a hidden treasure with audio plus transcriptions.

are broken, and the LM telemetry is unable to lock up. The noise on the air-to-ground communications loop is deafening. Controllers punched the loop off so they can hear communications among the flight control team.[14]

The noise lasted for two minutes. The 210-foot dishes at Goldstone, CA, and Madrid, Spain, were receiving weak signals from *Eagle*. Communications with the Command Module were solid, so Duke used Mike Collins as a relay.

> **DUKE:** *Columbia*, Houston. We've lost *Eagle* again. Have him try the high-gain.
> **COLLINS:** *Eagle*, this is *Columbia*. Houston lost you again. They're requesting another try at the high-gain.

Telecommunications Officer Don Puddy told Kranz, "I think this is probably due to some skin reflections."[15] Richard Stachurski, who was in the Network support room, later explained, "The steerable antenna, in its attempt to keep pointed at the Earth, may be pointing at or close to parts of the Lunar Module structure. The resulting reflections would distort the signal and account for the dropouts we are seeing."[16]

Apollo 12 Commander Pete Conrad walked up to the front of Charlie Duke's console and said, "How about letting them yaw?"[17]

Duke passed the idea on to Kranz: "Hey, Flight—suggestion— would a little yaw help this?" Kranz wanted Don Puddy's opinion.

Puddy replied, "Sure would, Flight." Most people would never think of starting an official communication in Mission Control with "Hey, Flight." The tight protocol was to give the call sign of the receiver first, followed by the call sign of the sender. But the descent was one continuous troubleshooting ordeal on the Flight Director loop, and Conrad and Duke's suggestion speaks volumes about the close teamwork within Mission Control.

---

[14] Kranz, *Failure Is Not an Option*, 284.

[15] Conversations on flight controller channels can be heard on the remarkable website Apolloinrealtime.org/11 assembled by Ben Feist with the help of archivist Stephen Slater.

[16] Richard Stachurski, *Below Tranquility Base, An Apollo Memoir*, (North Charleston, SC: CreateSpace Independent Publishing Platform, 2013), 247.

[17] Armstrong, Collins, and Aldrin, *First on the Moon*, 282.

Puddy checked with his back room and recommended *Eagle* rotate 10 degrees to the right. Kranz simultaneously summarized Puddy's statement and gave Duke the order to relay the message to the crew: "You want them to roll about 10 degrees, right?" Both parties quickly corrected him.

DUKE: It would be yaw, 10 degrees.
PUDDY: Yaw.
KRANZ: Yaw, Rog.

Kranz was thinking of piloting an aircraft where the thrust is horizontal, and a roll maneuver rotates the plane around its horizontal thrust axis. But the LM was peculiar. The control inputs were designed for the landing phase where it hovered over the Moon, balanced on a vertical thrust with its legs reaching toward the lunar surface. The pitch thrusters caused the legs to tilt forward and backward. Roll thrusters caused its legs to rock side to side. Yaw thrusters caused the LM to spin around its vertical thrust axis with its legs still pointing down. In the early stages of the descent, the LM was tilted back 90 degrees from its normal orientation with its legs pointed forward. The spin rotation Duke and Kranz referred to was accomplished by its yaw thrusters. The same maneuver in the CSM, which has a horizontal thrust axis, would have been performed with its roll thrusters.[18]

Duke called to *Eagle*: "We recommend you yaw 10 right. It will help us on the high-gain signal strength." He received no reply.

While Duke waited, Kranz polled his team of flight controllers to see if they were Go or No-Go for the landing attempt. All replied Go. Then the pesky communication problems flared up again. In simulations, the communications were either stable or completely off; the team never had to deal with intermittent dropouts.[19]

Duke called again, telling *Eagle* they were Go for powered descent. When he received no reply from the LM, Collins stepped in as a relay. "*Eagle*, this is *Columbia*. They just gave you a Go for powered descent."

---

[18] For a more detailed explanation, *see* George A. Zupp, "An Analysis and a Historical Review of the Apollo Program Lunar Module Touchdown Dynamics," (Houston: NASA Johnson Space Center, 2013), 7. https://www.lpi.usra.edu/lunar/documents/SP-2013-605.pdf.

[19] Stachurski, *Below Tranquility Base*, 246.

Charlie Duke wanted to make sure Collins relayed the previous message as well:

> **DUKE:** *Columbia*, Houston. We've lost them on the high-gain again. We're recommending they yaw right 10 degrees and reacquire.

While Duke was passing the message on to Collins, Don Puddy suggested to Kranz they try the rear of the two omnidirectional antennas. Trying to get Duke's attention, Kranz urgently called, "Break, break, CAPCOM. Break, break, CAPCOM. We'd like to go Aft Omni."

At that moment, Collins was still relaying the yaw recommendation to *Eagle*. Kranz, Puddy, and Duke decided to wait and see if the yaw maneuver worked before complicating things with a new message. It did, restoring communications from a quarter of a million miles away. Armstrong and Aldrin pressed on with the pre-burn checklist.

TWO MINUTES BEFORE PDI

Aldrin reported the sequence (movie) camera was coming on. He referred to the Maurer Data Acquisition Camera (DAC) designed to record engineering data. There was one DAC in the CM and one in the LM. The LM version was equipped with an 18mm wide-angle lens and was attached to brackets in the top left corner of Aldrin's triangular window. The DAC was pointed down and forward, so the camera and astronaut had an unobstructed view of the lunar surface.

The camera contained 130 feet of film in its magazine and had settings for one, six, twelve, and twenty-four frames per second. For the moonwalk, Aldrin would set it to one frame per second so the camera would record most of the action. The camera's standard rate was 24 fps, allowing a run time of 3.6 minutes. For the landing, he set it to six fps, providing enough run time to film the full descent.[20]

Armstrong sounded bored in the final seconds before ignition as he sighed and said, "Okay, nothing else is left to do here." Aldrin armed

---

[20] NASA, "Apollo 11 Lunar Photography, April 1970," accessed February 23, 2020, https://ntrs.nasa.gov/api/citations/19720010768/downloads/19720010768.pdf, 3. The flight plan called for Aldrin to set the camera to six fps for the descent.

the descent engine. In preparation, a small ullage burn by the thrusters settled the propellants to the bottom of the tank.[21] The computer flashed a request to the crew, asking if they wanted to begin the landing attempt. Aldrin pressed Proceed in confirmation. It was time.

The next 12 minutes would see Armstrong, Aldrin, and the flight controllers attempt something never done before. The journey into the unknown was so daunting that many in NASA did not think the first landing attempt would be successful. They thought it more likely Pete Conrad on Apollo 12, not Neil Armstrong, would take the historic first steps on the Moon. Armstrong knew what he was up against:

> The final descent to landing was far and away the most complex part of the flight. The unknowns were rampant. The systems in this mode had only been tested on Earth and never in the real environment. There were just a thousand things to worry about in the final descent. It was hardest for the system and it was hardest for the crews to complete that part of the flight successfully.[22]

Allan Klumpp and Don Eyles were two extremely interested observers in Cambridge, Massachusetts. "Klumpp had designed the guidance equations. I programmed them into the onboard guidance computer," said Eyles.[23] The harsh, unforgiving environment of space was about to test their years of hard work:

> The start of the landing, the moment when the engine would be lit and the guidance equation engaged, was called Powered Descent Initiation, or PDI. That three-letter combination would raise the hairs on your neck too, if it were the moment your software was going to take control of the spacecraft, two lives, a national goal, and if you like, human destiny.[24]

---

[21] In zero-gravity conditions, propellants dispersed throughout the tank. Thrusters were fired to slightly decrease the velocity forcing the fuel to the engine intake and producing a smooth burn.

[22] Neil Armstrong, interview by Stephen Ambrose and Douglas Brinkley, 86.

[23] Don Eyles, "Landing Apollo via Cambridge," MIT News, July 17, 2009. https://news.mit.edu/2009/apollo-eyles-0717.

[24] Eyles, *Sunburst and Luminary*, 24.

POWERED DESCENT INITIATION

"Ignition," reported Armstrong and Aldrin simultaneously. The descent engine began with 26 seconds of 10 percent thrust, when the computer would calculate the center of mass and point the thrust directly through it. Static immediately filled the air-to-ground loop, signaling the loss of communication and data. Charlie Duke clenched his teeth in frustration. Doug Ward knew the implications of the communication problems.

> During the landing phase, I was most worried about the communications dropouts because the Mission Rules said you had to have good data and voice communications during the landing. And the first thing that happened after PDI was that we started to get dropouts in communications and data. It was pretty serious, and I was very concerned that it was going to abort the landing. Mission Rules were really the backbone of mission operations. They were not really rules as much as guidelines. The Flight Director, and this was often misunderstood by the press, had a complete latitude to interpret those rules based on the current set of circumstances that they were facing, and they might not necessarily jibe with the conclusion the rules would appear to dictate.[25]

Fortunately, Kranz decided to press on. Charlie Duke asked Kranz and Puddy if they wanted to try their previous suggestion of the aft omnidirectional antenna. They concurred. Again, Duke used Collins to relay the information. "*Columbia*, Houston. We've lost them. Tell them to go aft omni, over."

The communication problems were annoying, but Steve Bales stared at a potential showstopper. At ignition, the LM was dropping toward the lunar surface 20 feet per second faster than expected. If the variation reached 35 feet per second, it would cross the acceptable limit, and Bales would have to call an abort.

There were two possible causes. Either *Eagle* was farther downrange than expected, or there was a problem with the guidance platform. Bales hoped it was the former since there was a correlation

---

[25] Doug Ward, interview by author, March 16, 2019.

between down-track position and the higher vertical velocity.[26] If that were the case, the error would remain constant until the landing radar achieved lock-on. Then the computer would gradually correct the discrepancy once the radar data was entered. However, if the extra descent rate were due to a failing guidance platform, the variance would increase, and the landing attempt would be aborted.

After 26 seconds, the descent engine throttled up to full power. Bales was still staring at the vertical velocity discrepancy, which remained steady at 20 fps. He reported to Kranz, "At turn on, we had a 20 foot per second residual that is probably due to down-track error." Kranz repeated the message back to Bales while recording it neatly in his logbook. Bales added, "and it was in radial," signifying the error was downward toward the Moon.

Kranz checked back with Bales a minute later to see how the data looked. Bales replied, "Hanging out 20 foot per second, looks good."

"Rog. No change is what you're saying."

"No change. That's down-track, and I know it."

Bales repeated it to FDO Jay Greene. "It's down-track, Jay, I know it. We'll be okay if it doesn't grow."

PAO (WARD): Two Minutes twenty seconds, everything looking good. We show altitude of about 47,000 feet.

2:20 INTO POWERED DESCENT

Kranz asked again about the descent velocity error. Bales replied, "Holding at 18 foot. We're going to make it, I think." Armstrong made his final downrange position check over the crater Maskelyne W, which passed by the window three seconds sooner than expected.[27] Armstrong reported on the air-to-ground loop, "Our position checks downrange show us to be a little long." It corroborated Bales' hunch and was good news to the young Guidance Officer. "We CONFIRM that," Bales told Kranz and Duke confidently.

The early problems in the descent helped the flight controllers settle their nerves and get into their rhythm but were agonizing for the men

---

[26] Steve Bales, interview by author, October 9, 2019.
[27] Armstrong, Collins, Aldrin, *First on the Moon,* 288.

sitting behind Gene Kranz, who could only watch and wait. On the back row of the MOCR sat NASA's greatly respected leadership: Chris Kraft, Bob Gilruth, George Low, and Sam Phillips.

Kraft remembers, "I couldn't stop myself from clenching the arms of my chair and from starting to breathe in shorter and shorter pulses."[28]

3:30 INTO POWERED DESCENT

Armstrong told Aldrin onboard, "About three seconds long. Rolling over."[29] *Eagle* rotated 180 degrees to aim its landing radar at the lunar surface while Gene Kranz polled his flight controllers for a Go/No-Go decision to proceed with the descent. When he got to Bales, Steve bellowed GO! Kranz chuckled as he continued.[30]

When *Eagle* completed its rotation, the crew faced away from the Moon. Aldrin took his eyes off the computer display momentarily, noticing a familiar sight. He said, "We've got the Earth right outside our window."

5:00 INTO POWERED DESCENT

The onboard guidance computer had been relying on data given to it by Greene and Bales two hours earlier. The landing radar was more accurate as it gave precise real-time measurements; thus, it was used to correct the onboard guidance system. As the radar engaged the lunar surface, it confirmed what Mission Control had already suspected. The landing radar showed the LM was 2,900 feet lower than it was supposed to be, a discrepancy referred to as "Delta-H." It also verified the 20 fps sink-rate error.

At this point, Buzz Aldrin was to enter two commands into the computer. The first was a particular program allowing him to monitor the convergence of the Delta-H. The convergence would indicate the

---

[28] Kraft, *Flight, My Life in Mission Control*, 320.

[29] At that point in the descent, hitting their landmarks three seconds early would result in a landing point about three miles downrange of the target.

[30] Five decades later, Kranz remembered the scene well. "Bales was pumped. The whole room heard him literally yelling his Go, and there were several snickers from my controllers." Gene Kranz, 2018 Vice Adm. Donald D. Engen Flight Jacket Night Lecture.

computer was working well. If so, he would enter another command telling the computer to accept the landing radar's information as accurate. The computer would then incorporate the data. An algorithm gradually corrected its information to match the radar.[31]

A computer program called P63 ran the initial part of the descent, the "Braking Phase." P63 used 85 percent of the computer's capacity, allowing a margin for additional activity if needed. The special program Aldrin typed in to monitor the Delta-H added three percent more drain on the system, which should have been fine. But the computer wasn't fine. The program threw the computer over its capacity to do all requested jobs in its two-second cycles. The computer set off an alarm, rebooted instantaneously, and continued with only the highest priority responsibilities. The alarm sent shockwaves not only to the crew but also to the flight controllers a quarter of a million miles away.

Even though Jack Garman was in a back room in Mission Control, he could imagine what was going on in *Eagle's* cockpit. "In the vehicle, when one of these alarms came up, it would ring what was called the master caution and warning system. Now, master caution and warning is like having a fire alarm go off in a closet. I mean, 'I want to make sure you're awake,' one of these in the earphones, lights, everything. I gather their heart rates went way up. You're not looking out the window anymore."[32]

1202

Steve Bales was listening to four different communication loops simultaneously. Two heart-stopping words from *Eagle* pierced through the other conversations. "Program Alarm." Armstrong, then Aldrin, specified it was a 1202. Bales was shocked! He thought those alarms would never show up in a real mission. Neither did Don Eyles, who coded the software for the landing phase. Eyles had the same reaction as he listened from Cambridge. That alarm couldn't happen! But it did. "At

[31] Steve Bales, interview by author, October 9, 2019.
[32] Jack Garman, interview by Kevin Rusnak for NASA JSC Oral History, March 27, 2001, 24.

MIT, where we realized something mysterious was draining time from the computer, we were barely breathing."[33]

Eagle reported the alarm a few seconds before it appeared on Bales' display. He immediately called Jack Garman in the support room, sitting beside MIT software expert Russ Larson. In front of Garman was his program alarm cheat sheet based on conversations with MIT after the final simulation. Those hours of research over something they initially considered a waste of time were about to pay huge dividends. Steve Bales and his boss Jerry Bostick scrambled to find their copies, but Garman accessed his first.

Garman calmly told Bales, "It's an executive overflow. If it does not occur again we're fine." The computer was asked to do too many tasks. As designed, it flushed all of its duties and restarted, which only took a few seconds. Bales could see the rebooted computer was handling its primary responsibilities well. Twenty-two seconds after the program alarm occurred, Armstrong's voice penetrated the air-to-ground loop again. "Give us a reading on the 1202 program alarm."

Neil Armstrong did not speak often during the simulations; Aldrin handled most of the communications with the ground. The fact Armstrong was asking for a status report let Bales know how intense the situation was in the cockpit.[34]

Charlie Duke, listening intently to Bales and Kranz, was apprehensive. He recalled, "When I heard Neil say '1202' the first time, I tell you, my heart hit the floor."[35] He remembered the final simulation where similar computer alarms caused an abort.

It was decision time. Bales said, "We... we're Go on that, Flight." Kranz noted the stammer and repeated the statement to Bales with raised inflection. It was his way of asking, "Are you really sure about that?" Duke knew the crew needed a response immediately and did not wait for Bales to answer. He quickly told Neil and Buzz, "We're Go on that alarm." Bales recalls,

Thank goodness for Charlie. There was an ironclad rule in Mission Control that the Capsule Communicator does not say

[33] Eyles, "Tales from the Lunar Module Guidance Computer."
[34] Steve Bales, interview by author, October 9, 2019.
[35] Armstrong, Collins, Aldrin, *First on the Moon*, 287–288.

anything to the crew unless the Flight Director approves it. Charlie knew he'd better not wait. I think it's the only time during the descent or in simulations that a CAPCOM ever told the crew something without the Flight Director telling him it was okay. He took a risk because if Gene would have overridden me, Charlie would have been stuck with both feet in midair! He and I both would have had both feet in midair!"[36]

Bales replied to Kranz, "If it doesn't reoccur, we'll be Go."

Garman then clarified to Bales that the frequency of the alarms was what mattered. "It's *continuous* that makes it No-Go. If it reoccurs, we're fine." Russ Larson from MIT was so anxious he could not speak. He simply gave Garman a thumbs up.[37]

The tense thirty seconds following the program alarm should forever disprove the common female notion that men cannot multitask. Onboard *Eagle*, Armstrong and Aldrin were discussing the program alarm as well as the computer incorporating the landing radar's altitude discrepancy, Delta-H. Kranz was having the same two conversations with Bales. Meanwhile, Bales was talking to Garman in the back room. Amid those discussions, Retrofire Officer Chuck Deiterich reported to Kranz a new piece of critical information the crew needed: the time when the descent engine would throttle down due to the decreasing weight of the LM. It was akin to scenarios Dick Koos, Jay Honeycutt, and the simulations team concocted to train the team for such a time as this—create havoc and then pile on more chaos.

During the commotion, Armstrong and Aldrin told the computer to incorporate the landing radar data. After Duke reported the time they should expect the engine to throttle down, Aldrin again keyed in the command telling the onboard computer to display the convergence of the altitude numbers. Immediately the alarm reoccurred. It was now clear the alarms were related to the extra program monitoring the Delta-H.

Garman saw the second alarm and said to Bales, "Same alarm. Tell him to leave it alone, and we'll monitor it."

---

[36] Steve Bales, interview by author, October 9, 2019.
[37] Eyles, "Tales from the Lunar Module Guidance Computer."

Bales relayed to Kranz, "We'll monitor his Delta-H, Flight. And his alarms." Duke passed that information to the crew so they could focus on the descent.

Chris Kraft looked over the top of his console and watched Steve Bales on the front row. "Good kid," he said to Bob Gilruth and George Low. Then he asked them, "Were any of us ever that young?"[38]

Kranz told his team, "Okay all flight controllers, hang tight. We should be throttling down shortly."[39]

The sudden reduction from full thrust to 60 percent was unmistakable. Aldrin said, "Wow! Throttle down. You can feel it in here when it throttles down, better than in the simulator."

Eyles was breathing again. Throttling down on time meant the guidance computer was performing its main tasks.[40] They were halfway through the landing phase.

PAO (WARD): Altitude now 21,000 feet. Still looking very good. Velocity down now to 1,200 feet per second.

7:00 INTO POWERED DESCENT

Duke updated the crew, "At seven minutes, you're looking great to us, *Eagle*." The LM was gradually pitching over as it fell from orbit. Soon the crew would be able to see the lunar surface again.

A confident Bales reported to Kranz at seven and a half minutes, "His landing radar is fixed, and the velocity is beautiful."

Bob Carlton, overseeing the LM fuel supply, told Kranz the next piece of information the crew needed to know, "Descent 2, fuel." The LM propellant tanks had two independent methods for measuring the remaining quantity. During the descent, flight controllers judged the second was the most reliable.

Charlie Duke relayed the information to Armstrong and Aldrin. "*Eagle*, Houston. It's Descent 2 fuel to monitor. Over."

---

[38] Kraft, *Flight, My Life in Mission Control*, 321.
[39] Excessive corrosion occurred when the thrust was between 65 percent and 93 percent of maximum, so the engine used full thrust until its decreased weight required less than 60 percent. *See* chapter 14, page 168.
[40] Eyles, *Sunburst and Luminary*, 27.

A reassured Steve Bales told Greene, "Jay, we're in good shape now, babe."

Greene reported to Kranz, "Flight, FDO, looking real good."

8:00 INTO POWERED DESCENT

Aldrin asked Houston for the estimated transition time to the next part of the descent, the "Approach Phase." It was controlled by a different computer program, P64, also coded by Don Eyles. When the onboard computer automatically switched to P64, the LM would pitch over to the point where Armstrong could see the landing site out his window. Bales quickly responded to Aldrin's request, "30 seconds to P64."

While Duke passed the information along, Kranz asked if they still had landing radar. Bales confirmed and said it had converged with the onboard guidance system.

Jay Greene updated Kranz again. "Flight, FDO, We're Go, looks real good."

Duke heard Greene. "*Eagle*, Houston, coming up on 8:30. You're looking great."

Armstrong reported P64 occurred right on time. This was another good sign for Don Eyles; the computer was still working properly despite the two program alarms. But something was mysteriously using up about 15 percent of the computer's resources.

Eagle's altitude was down to 5,000 feet, descending at 100 fps. Once they were in P64, Armstrong briefly disengaged the Lunar Module's autopilot to check his manual controls in pitch and yaw. When they were responsive, Armstrong reported, "Manual attitude control is good." He put *Eagle* back on autopilot as the descent continued.[41]

PAO (WARD): We're now in the approach phase, everything looking good.

Kranz polled all the flight controllers for a Go/No-Go decision for landing at nine minutes. The controllers sounded confident. Kranz told Duke to pass along the result.

---

[41] NASA, "Apollo 11 Technical Crew Debriefing," 9-23.

DUKE: *Eagle,* Houston. You're Go for landing. Over.
ALDRIN: Roger. Understand. Go for landing, 3,000 feet.

Chris Kraft had been waiting to hear those words. "Go for landing! I knew that every heart and every set of lungs in Mission Control—and maybe around the world—stopped in that second and waited."[42]

P64 used up about 90 percent of the computer's capacity. The computer had less work to do when Armstrong put it in manual control, but when he put it back on autopilot, the P64 program plus the mystery bug threw it over the cliff again.[43] Aldrin, then Armstrong, reported another program alarm. This time it was a 1201.[44]

Jack Garman immediately told Bales, "Same type, we're Go." Bales relayed it on to Kranz and Duke.

Charlie Duke quickly passed it up to the crew, "We're Go. Same type. We're Go." Garman thought it was funny to hear his words echo from Steve Bales to Charlie Duke out to the rest of the world, including Neil Armstrong and Buzz Aldrin a quarter of a million miles away.[45]

9:30 INTO POWERED DESCENT

The computer display went blank for about 10 seconds after the 1201 alarm.[46] After the flight, Armstrong was asked what he thought about the display going dark. "Seemed like a long time," Armstrong chuckled, "I never expected it to come back."[47] Armstrong explained the problems caused by the alarms:

> Normally in this time period, that is, from P64 onward, we would be evaluating the landing site and checking our position. However, the concern here was not with the landing area we were going into, but rather whether we could continue at all. Consequently, our attention was directed toward clearing the program alarms, keeping the machine flying, and

---

[42] Kraft, *Flight, My Life in Mission Control,* 321.
[43] Steve Bales, interview by author, October 9, 2019.
[44] This was a similar alarm and would be handled the same way as earlier.
[45] Jack Garman, interview by Kevin Rusnak for NASA JSC Oral History, March 27, 2001, 25.
[46] Steve Bales, interview by author, October 9, 2019.
[47] Eyles, *Sunburst and Luminary,* 163.

assuring ourselves that control was adequate to continue without requiring an abort. Most of the attention was directed inside the cockpit during this time period, and in my view, this would account for our inability to study the landing site and final landing location during the final descent. It wasn't until we got below 2,000 feet that we were actually able to look out and view the landing area.[48]

After clearing the program alarm, Neil Armstrong could finally get his eyes out of the cockpit and assess where the autopilot was taking them. He asked Aldrin for the Landing Point Designator (LPD), a number on the computer display giving an angle indicating where the LM was targeting. On the Commander's forward window was a vertical line etched into the inner and outer panes. He positioned himself so the two lines were superimposed. On the line were numbered hash marks. Armstrong found the number corresponding to the LPD reading, which showed him where the LM was heading. At first glance, he said, "That's not a bad-looking area." Armstrong liked what he saw:

It was indicating we were landing just short of a large rocky crater surrounded by a large boulder field with very large rocks covering a high percentage of the surface. I initially felt that might be a good landing area if we could stop short of that crater, because it would have more scientific value to be close to a large crater.[49]

Another program alarm occurred. Duke heard "Roger, no sweat" on the Guidance loop. He quickly told the crew, "1202. We copy it." It was his shorthand way of saying, "Keep doing what you're doing, and we'll let you know if there is anything to worry about." Aldrin repeatedly called out the altitude, descent rate, and LPD angles.

Kranz quickly polled Carlton, Puddy, Bales, and Greene, the four controllers watching the most critical data as the LM approached the surface. As they finished telling Kranz everything looked good, ANOTHER 1202 alarm sounded.

---

[48] NASA, "Apollo 11 Technical Crew Debriefing," 9-21.
[49] NASA, "Apollo 11 Technical Crew Debriefing," 9-23.

Armstrong focused on finding a suitable landing area while Aldrin called out the data. The remaining fuel sloshed in the tank, rocking the LM and causing the LPD readings to vary. Aldrin frequently updated Armstrong.

> ALDRIN: Seven hundred feet, 21 down, 33 degrees. [Altitude, rate of descent, LPD indicator]
> ARMSTRONG: Pretty rocky area.

Unknown to the flight controllers, when Armstrong checked the 33-degree mark on the Landing Point Designator, he realized the LM was headed toward a crater the size of the Houston Astrodome.[50] On the landing charts, it was called "West Crater" since it was on the western edge of the landing ellipse. Huge boulders surrounded its raised rim. Armstrong recalled, "Continuing to monitor LPD, it became obvious that I could not stop short enough to find a safe landing area."[51]

Under 600 feet, Armstrong took control of the forward velocity with a hand controller and used a toggle switch with his left hand to increase or decrease the rate of descent. A better area appeared half a mile in the distance, and he hurried forward over West Crater.

Deke Slayton was sitting to Duke's right. They realized Armstrong and Aldrin had their hands full and did not need to be interrupted. Duke told Kranz, "I think we'd better be quiet, Flight." Kranz agreed and said, "The only callouts from now on will be fuel."

> ALDRIN: Four hundred feet. Down at nine, 58 forward.

The 58-feet-per-second forward velocity raised many eyebrows in Mission Control. By this point in simulations, the LM was typically settling onto the lunar surface.

10:30 INTO POWERED DESCENT

Eagle was in its Landing Phase, run by Program 66. Flight controllers noted the change to P66 amongst themselves but did not say anything

---

[50] The Astrodome was the first indoor multi-purpose stadium.
[51] NASA, "Apollo 11 Technical Crew Debriefing," 9-23.

on the Flight Director loop. Bales was acutely aware of the situation. "Bob Carlton was making the critical fuel calls, and it would have been terrible if someone had voiced over him at an important time. At that altitude, no one other than Bob had an abort call."[52]

At 330 feet, Aldrin said, "We're pegged on horizontal velocity." Houston never had a sim like this. Armstrong skimmed westward over the lunar surface, searching for a place to land. The critical issue was no longer the computer but the quantity of propellant.

Armstrong asked, "How's the fuel?"

"Eight percent," replied Aldrin.

Armstrong picked out his landing spot beyond an upcoming crater. "Looks like a good area here."

11:00 INTO POWERED DESCENT

Kranz was thinking about the fuel as well. He alerted Carlton, who was monitoring the fuel level. "Okay, Bob, I'll be standing by for your callout shortly."

As *Eagle* approached what would later be known as Little West Crater, the landing radar briefly lost its lock with the surface. Bales commented on the Guidance loop, "Looks like radar dropped out."

Armstrong rocked back to slow *Eagle's* forward velocity. At 200 feet, Armstrong calmly told Aldrin the landing site would be "right over that crater."

11:30 INTO POWERED DESCENT

Eagle drifted 120 feet above Little West Crater while Bob Carlton called out "Low Level" and started his Heuer stopwatch.

In Bill Tindall's May 29, 1969 Missions Techniques meeting, there was a comprehensive review of the low-level propellant light. Tindall wrote, "The most significant piece of information coming from this was that we are assured of about 98 seconds more descent engine operation at the hover thrust level after the light comes on."[53] In simulations, flight

---

[52] Steve Bales, email message to author, November 6, 2019.
[53] Bill Tindall, "DPS low-level propellant light," Tindallgram, May 29, 1969, http://www.collectspace.com/resources/tindallgrams/tindallgrams02.pdf.

controllers had never seen the crew go this long before landing. They began holding their breath.

Decades later, Carlton told the BBC, "When I started my stopwatch, I didn't think there was a chance in the world of us landing. I was looking at the altimeter, and we had a heck of a way to go." He revealed how he had prepared for this moment:

> I put a little piece of Scotch tape—a very scientific thing here now—a piece of Scotch tape with a mark on it that said when you hit this one there are 60 seconds left. And when it comes on down to this one, there are 30 seconds left. And then when it comes down to the point where we need to abort, we don't want to be empty at that time. There's some uncertainty in this measuring process, so we need to put a margin in there to be sure we don't run into that uncertainty. We put a little extra in there to account for that.[54]

One hundred feet above the lunar surface, Aldrin advised Armstrong they had five percent fuel left. The low-level warning light radiated in the cockpit.

11:50 INTO POWERED DESCENT

The second hand on Carlton's stopwatch hit the 60-second mark, and he informed Kranz, who repeated it. Duke reported it to the crew. Armstrong had 60 seconds until he would hear an abort call. "And now you've got this multibillion-dollar program, and it's riding on Neil and Buzz and Bob Carlton and his stopwatch," Bales remembers. "It's really riding on Neil, who might have overridden us if we said no fuel."[55]

> ALDRIN: Forty feet, down two and a half. Picking up some dust.

---

[54] Bob Carlton's stopwatch is now on display at the Smithsonian National Air and Space Museum in Washington, DC. "Thirteen Minutes to The Moon," Season 1, Ep. 09, "Tranquility Base," BBC, July 10, 2019, https://www.bbc.co.uk/programmes/w3csz4ds.
[55] Steve Bales, interview by author, October 9, 2019.

Dust! Hearing that comment was "the most phenomenal point" for Jack Garman. "We'd watched hundreds of landings in simulation, and they're very real, and on this particular one, the real one, the first one, Buzz Aldrin called out, "We've got dust now," and we'd never heard that before. You know, it's one of those, 'Oh, this is the real thing, isn't it?'"[56]

The dust created problems for Armstrong. A transparent sheet of particles blew away from the LM in all directions, and the visibility deteriorated as their altitude decreased. He struggled to determine whether he was drifting forward, backward, or sideways. "I found that to be quite difficult."[57]

ALDRIN: Thirty feet, two and a half down. Faint shadow. Four forward, four forward. Drifting to the right a little.

12:20 INTO POWERED DESCENT

Carlton's eyes were glued to the stopwatch. He called out "30 seconds," and Duke relayed it to the crew. *Eagle* started drifting left and backward. In the post-flight debriefing, Armstrong said, "That's the thing I certainly didn't want to do, because you don't like to be going backward, unable to see where you're going. So I arrested this backward rate with some possibly spastic control motions, but I was unable to stop the left translational rate."[58]

Visibility worsened as they approached the surface. "The blowing dust became increasingly thicker. It was very much like landing in a fast-moving ground fog," said Armstrong.[59]

Sensing probes extended five and a half feet below Eagle's footpads. [60] After what seemed like an eternity, one of the probes touched the lunar surface and a glorious blue contact light gleamed in the cabin. This alerted the crew to shut off the engine before landing to prevent a boulder from obstructing the engine nozzle while running.

---

[56] Jack Garman, interview by Kevin Rusnak, p.25.
[57] NASA, "Apollo 11 Technical Crew Debriefing," 9-24.
[58] NASA, "Apollo 11 Technical Crew Debriefing," 9-25.
[59] Armstrong, Collins, Aldrin, *First on the Moon,* 292.
[60] Bennett, "Apollo Experience Report," 8–11.

With only a hint of excitement, Aldrin said, "Contact light." As *Eagle* nestled onto the lunar surface, Armstrong reported engine shutdown. Aldrin announced to the world, "Okay, engine stop."

Jay Greene calmly said, "Touchdown," as if he was reading a phone book. Charlie Duke exhaled with relief.

Neil Armstrong could not believe what happened outside the window:

> I was absolutely dumbfounded when I shut the rocket engine off and the particles that were going out radially from the bottom of the engine fell all the way out over the horizon, and when I shut the engine off, they just raced out over the horizon and instantaneously disappeared, you know, just like it had been shut off for a week. That was remarkable. I'd never seen anything like that. And logic says, yes, that's the way it ought to be there, but I hadn't thought about it and I was surprised.[61]

Armstrong and Aldrin, consummate engineers, quickly ran through a list of well-rehearsed post-landing procedures to "safe" the Lunar Module. This included deactivating the thrusters. If they landed on a slope, the thrusters may have fired to keep the LM level which could tip over the spacecraft.

Aldrin finished, saying, "413 is in," an entry telling the abort guidance system they were on the Moon and it could reset its altitude to zero and its velocity to the lunar rate. Therefore, the ascent stage would have a good idea of its location if they had to leave the surface immediately and rendezvous with Collins.[62]

Dry-witted Gran Paules, sitting beside Bales, said, "Fantastic, we could do it every day."

Charlie Duke recalls, "Then the coolest guy in the world that I ever knew as a pilot, Neil Armstrong, said, 'Houston, Tranquility Base here. The *Eagle* has landed.' Just sort of nonchalant."[63] The statement surprised Steve Bales. Armstrong always used the call sign *Eagle* after

---

[61] Neil Armstrong, interview by Stephen Ambrose and Douglas Brinkley, 84.
[62] Steve Bales, interview by author, October 9, 2019.
[63] Charlie Duke, Message to Southside Community Church, Corpus Christi, Texas, September 6, 2018.

landing in the simulations. When Bales heard "Tranquility Base," he thought, "What a great name!"[64]

Armstrong did not want to catch CAPCOM off guard with the new terminology. In a meeting before the launch, Armstrong told Duke, "I'm going to change the call sign to Tranquility Base." Duke said, "There were four of us in on that, myself and the three crewmen. It surprised a lot of people. I knew it was coming, but I was so excited that I couldn't even pronounce Tranquility."[65]

A tongue-tied Duke responded, "Roger, Twan—Tranquility. We copy you on the ground. You got a bunch of guys about to turn blue. We're breathing again. Thanks a lot."

Armstrong and Aldrin looked at each other and shook their gloved hands. Hard. Armstrong later told Honeysuckle Creek's Hamish Lindsay, "If there was an emotional high point, it was the point after touchdown when Buzz and I shook hands without saying a word. That still, in my mind, is the high point."[66]

Aldrin writes, "We had so much to do, and so little time in which to do it, that we no sooner landed than we were preparing to leave, in the event of an emergency. I'm surprised, in retrospect, that we even took time to slap each other on the shoulders."[67]

Eagle on the Moon with Earth overhead.
Photo credit: NASA AS11-40-5924.

---

[64] "13 Minutes to the Moon," Season 1, Ep. 09, BBC.

[65] Charlie Duke, interview by author, July 3, 2020.

[66] Hamish Lindsay, "Apollo 11, 16–25 July, 1969," accessed November 9, 2019, https://www.honeysucklecreek.net/msfn_missions/Apollo_11_mission/index.html.

[67] Colonel Edwin E. "Buzz" Aldrin, Jr., *Return to Earth*, with Wayne Warga, (New York: Random House, 1973), 232.

CAPCOM Charlie Duke during the tense Apollo 11 lunar landing. To his left are backup crew members Jim Lovell and Fred Haise, who flew on Apollo 13. Screen capture by Colin Mackellar from NASA 16mm footage.

Jack Garman (right) receiving an award from Chris Kraft. Photo credit: NASA.

APPLICABLE TO: IN DESCENT, AVERAGE-G ON

ALARM CODE	TYPE	PRE-MANUAL capability	MANUAL capability
.0105 MK ROUT. BUSY	POODOO		
00430 CAN'T INTG. 3V.	"		PGNCS GUIDANCE N₀/GO
01103 CCSHOLE- PROG. BUG	"	PGNCS GUID. LOST,	(PGNCS GO for
01209 NEG. WAITLIST	"	PGNCS/AGS BEST FOOT DN	TAPE METERS, CROSS-POINTERS,
01206 DSKY, TWO USERS	"		CONTROL,
01302 NEG. SQ. ROOT	"	(decision how on	ABORTING)
01501 DSKY, PROG. BAD	"	current rules)	
01502 DSKY, PROG. BUG	"	(NO LR DATA)	(NO LR DATA)
00607 LANE, NO SOLN	"		
"O.F." = Overflow too many.		DUTY CYCLE MAY DEGRADE	
CONTINUING →		PGNCS (AGS CONTROL MAY	
OCCURRENCE OF:		HELP-SEE BELOW)	SAME AS LEFT
01104 DELAY WT. O.V.	BAILOUT	(WATCH FOR OTHER CUES)	(except "other cues"
01201 EXEC. O.F. (VAC)	"	PGNCS CONTROL UNCERTAIN	which would otherwise
01202 EXEC. O.F. (Jobs)	"	DSKY MAY BE LOCKED UP,	be cause for ABORT
01203 EXEC. O.F. (Task)	"	DUTY CYCLE MAY BE UP	PROBABLY ACT'N,
01207 EXEC. O.F. (Irpt)	"	TO POINT OF MISSING SOME	INSTEAD IT WOULD
01210 TWO USERS	"	FUNCTIONS (THEY LEFT TO D.C.	BE PGNCS GUIDANCE
01211 MRK ROUT. INTRPT	"	SWITCH TO AGS (FOLLOW FDO	NO GO — COMPLETE MANUAL
02000 DAP O.F.	"	NEEDED) MAY HELP PRODUCE.	CONTROL IN DESC.)
		PGNCS DUTY CYCLE DEGRADES	
ISS WARNINGS WITH:			
00777 PIPA FAIL	LIGHT ONLY		
03777 CDU FAIL	"	PIPA/CDU/IMU FAIL	
04777 PIPA, CDU FAIL	"	DISCRETES PRESENT	same as left
07777 IMU FAIL	"		
10777 PIPA, IMU FAIL	"	(Other mission rules	
13777 CDU, IMU FAIL	"	suffice; alarm may help	
14777 PIPA, CDU, IMU FL	"	point to what rule will	
		be broken)	
00214 IMU TURNED OFF	LIGHT ONLY	AGS ABRT/ABRT STAGE	SWITCH TO AGS
			PGNCS NO/GO on G or C
			(poss. NO/GO on NAV.)
01107 E-MEM. Restored	FRESH STRT	AGS ABRT/ABRT STAGE	SWITCH TO AGS,
			PGNCS NO/GO;
			(IMU as ref. okay)
CONTINUING →		IG ALARM DOESN'T STOP:	IF ALARM DOESN'T STOP:
00402 BAD GUID. CMDS	LIGHT ONLY	same as "Poodoo's" ("next"?)	Same as "PooDs"
CONTINUING →			
01406 GUID. NO SOLN	LIGHT ONLY	PGNCS GUID. NO/GO AS LONG	same as left
01410 GUID. O.V.	LIGHT ONLY	AS ALARM OCCURRING	(except prob. as above)
		(ATT. HOLD, CONST. GTC, CONT. IK)	
		(ABRT WILL PROB. COME FROM	
		CURRENT RULES e.g. GTC vs. V)	
		WATCH GTC ←	

Jack Garman's cheat sheet which guided him and Steve Bales through the computer alarm crises which threatened the first lunar landing. Bales recalls, "I signed it about a week before the launch; it confirmed I had reviewed the material and accepted its conclusions."[68]
Credits: Jack Garman and Steve Bales.

---

[68] Steve Bales, email message to author, December 13, 2022.

Bob Carlton's stopwatch at the National Air and Space Museum.
Photo credit: National Air and Space Museum.

Charlie Duke breathing again after touchdown
and smiling at cameraman Jerry Bray. Screen capture
by Colin Mackellar from NASA 16mm footage.

Neil Armstrong in the Lunar Module simulator.
Photo credit: NASA S69-38677, Kipp Teague.

20

A LESSON FOR SOME OF US LIKE ME

STAY/NO-STAY

After a brief moment of jubilation in Houston, the flight controllers had to ensure the Lunar Module was stable and could remain on the Moon. Certain conditions would require *Eagle* to leave the lunar surface immediately. Armstrong said, "There were going to be all kinds of conceivable difficulties with plumbing, and valves, and pressure systems, relief valves, and so on. So we were ready to leave if we had to, and we were listening carefully to their instructions."[1]

The crew and flight controllers had to decide periodically if they should press on with the mission or depart for a rendezvous with the Command Module. The first decision had to be made one minute after the landing.

Once again, the wisdom of Bill Tindall was on display. In typical NASA lingo, "Go" means to proceed as planned. But Tindall realized the potential confusion regarding this phrase in a decision to "stay" on the Moon after landing. Four months before the launch of Apollo 11, Tindall brought it up in a meeting with his Mission Techniques group. His resulting Tindallgram summarized the discussion:

> It deals with terminology—specifically, use of the expression "Go/No-Go" regarding the decision whether to stay or abort

---

[1] Neil Armstrong, interview by Stephen Ambrose and Douglas Brinkley, 102.

immediately after landing on the lunar surface. Every time we talk about this activity, we have to redefine what we mean by "Go" and "No-Go." That is—confusion inevitably arises since "Go" means "stay" and "No-Go" means "abort" or "go." Accordingly, we are suggesting that the terminology for this particular decision be changed from "Go/No-Go" to "Stay/No-Stay" or something like that. Just call me "Aunt Emma."[2]

Steve Bales still remembers the meeting:

There is a lesson there for some of us like me. When we first heard that, Jay Greene and I looked at each other and said, "Bill Tindall knows a lot of stuff, but he's not a flight controller. He's not the one who should be saying this. He doesn't know the detailed nitty-gritties of what we did in the Control Center every day on the loops. That's not his call." Well, I don't care how much you know about something; you'd better pay attention to common sense. And common sense says Bill was right.[3]

After the meeting, Tindall talked to Gene Kranz about the Stay/No-Stay terminology. Kranz loved the idea. Bales continued, "The primary rule of Mission Control is that you don't use words incorrectly. You have to be precise. So that was a lesson for us. Even though you think you know what you're doing, sometimes you'd better listen to someone from another perspective because they just might be right."[4]

Bales recalls the Tindallgram which emerged after that 1969 meeting. "Everybody read it, and everybody said, 'Yeah, that's the way we should do it—except Jay and I!'"[5]

Aunt Emma won the day, and the first Stay/No-Stay decision was the immediate focus after the landing. As Charlie Duke talked about a bunch of guys about to turn blue, Jay Greene's backroom remained hard at work: "FDO, it still looks good." Brushing aside one of the most

---

[2] Bill Tindall, "G Lunar Surface stuff is still incomplete," Tindallgram, March 7, 1969, http://www.collectspace.com/resources/tindallgrams/tindallgrams02.pdf.
[3] Steve Bales, interview by author, October 9, 2019.
[4] Steve Bales, interview by author, October 9, 2019.
[5] Steve Bales, interview by author, October 9, 2019.

significant events in human history, Greene simply said, "Rog. Stay with it."

Bales and Greene discussed the condition of the LM, which looked great. Controllers made the first Stay decision a minute after the landing. They made the next one five minutes later. The final Stay decision would occur in two hours when Mike Collins orbited overhead on the next revolution.

Chris Kraft walked from the back row of Mission Control to the front row where Bales and Greene sat. He stood behind Bales, patting him on the shoulders and smiling like a proud papa. The team had come a long way since the first powered descent simulation several months earlier when Kraft plugged in beside Bales. Kraft moved over to Greene and congratulated him as well. Few words were needed.

Steve Bales was busy for several minutes following touchdown. Then he reflected with Jack Garman on the significance of their conversation about the program alarms. He called Garman on the loop. "Hey Jack, thank God we had that meeting!"

Armstrong took a few moments to explain why he had to search so long for a landing spot:

> Hey, Houston, that may have seemed like a very long final phase. The auto-targeting was taking us right into a football field-sized crater, with a large number of big boulders and rocks for about one or two crater diameters around it, and it required us going in P66 and flying manually over the rock field to find a reasonably good area.

After the flight, Armstrong told mission planner Floyd Bennett, "I was just absolutely adamant about my God-given right to be wishy-washy about where I was going to land."[6]

Charlie Duke updated the men at Tranquility Base:

> **DUKE:** Be advised there are lots of smiling faces in this room and all over the world.
> **ARMSTRONG:** Well, there are two of them up here.

---

[6] Floyd V. Bennett, interview by Jennifer Ross-Nazzal for NASA Johnson Space Center Oral History Project, October 22, 2003, 9.

COLLINS: And don't forget one in the Command Module. Tranquility Base, it sure sounded great from up here. You guys did a fantastic job.
ARMSTRONG: Thank you. Just keep that orbiting base ready for us up there.

Next, Armstrong brought up a question Mission Control was pondering as well:

ARMSTRONG: Houston, the guys that said we wouldn't be able to tell precisely where we land are the winners today. We were a little busy worrying about program alarms and things like that in the part of the descent where we would normally be picking out our landing spot. And aside from a good look at several of the craters we came over in the final descent, I haven't been able to pick out the things on the horizon as a reference yet.
DUKE: Rog, Tranquility. No sweat. We'll figure it out.

In the Science Support Room at Mission Control, Dr. Gene Shoemaker of the US Geological Survey headed up a team of eight geologists who pored over their maps, listening for clues indicating *Eagle's* location. Four others did the same in a nearby building.

Twenty minutes after landing, Armstrong mentioned, "The area out the left-hand window is a relatively level plain cratered with a fairly large number of craters of the five to fifty-foot variety." Shoemaker said:

Tracking information indicated that Neil had flown past the middle of the landing ellipse and was several kilometers downrange. We knew that he had flown over a blocky-rim crater, that he had seen rays of ejecta as he passed over, and that the landing pattern had been rather like a fish hook. There were maybe six craters which could fit his description, but once we knew he was downrange, we narrowed it to two. I believe it was Marita West, over in Building Number 2, who first suggested that the crater Neil had described was West Crater. All of us came to the same conclusion pretty rapidly.[7]

---

[7] Armstrong, Collins, Aldrin, *First on the Moon,* 248.

Many on Earth were looking up in amazement at the Moon. Armstrong was also looking up, admiring his home planet. He told Duke, "From the surface, we cannot see any stars out the window, but out my overhead hatch I'm looking at the Earth. It's big and bright and beautiful. Buzz is going to give a try at seeing some stars through the optics."

Duke said, "Roger, Tranquility. We understand. It must be a beautiful sight."

At 4:00 p.m., Collins went around the western limb of the Moon. Duke informed him, "You're looking great going over the hill." Then Collins was out of contact with the rest of humanity. He reflected on the experience:

> I don't mean to deny a feeling of solitude. It is there. I am alone now. Truly alone. And absolutely isolated from any known life. I am it. If a count were taken, the score would be three billion plus two over on the other side of the Moon, and one plus God only knows what on this side. I feel this powerfully, not as fear or loneliness, but as awareness, anticipation, satisfaction, confidence, almost exaltation. I like the feeling.[8]

A few hours after touchdown, it was clear *Eagle* was stable and would not need to leave the lunar surface earlier than planned. The crew received their third and final directive to Stay. They ceased the simulated countdown, which prepared the crew in case an emergency launch was needed. The simulated countdown also served as practice for the actual lunar liftoff the following day. If a problem occurred, the ground would have extra time to address it.

Kranz's White Team began to hand over Mission Control to Milt Windler's Maroon Team while the astronauts prepared for dinner and their rest period. It had been a long day, and the moonwalk was scheduled for the next morning.

ALDRIN: Houston, Tranquility Base is ready to go through the power down and terminate the simulated countdown.

---

[8] Collins, *Carrying the Fire*, 402.

DUKE: You can start your power down now. And, Tranquility Base, the White Team is going off now and letting the Maroon Team take over. We appreciate the great show. It was a beautiful job, you guys.
ALDRIN: Roger. We couldn't have had better treatment from all of you back there.

Armstrong had one final request before Duke left. "Our recommendation at this point is planning an EVA with your concurrence, starting about 8:00 this evening, Houston time. That is about three hours from now."

Duke said, "Stand by."

Armstrong said, "Well, we will give you some time to think about that."

It only took a few seconds before he had the reply. Duke said, "Tranquility Base, Houston. We thought about it; we will support it. We're Go at that time. You guys are getting prime time TV there."

"I hope that little TV set works, but we'll see."

Moving up the EVA was a surprise to nobody except possibly Bruce McCandless, the EVA CAPCOM. Knowing the flight plan called for a rest period before the moonwalk, he drove home after the landing to eat dinner. "As I turned into my driveway, my wife came running out, waving her arms and shouting, 'Go Back!'" said McCandless.[9]

Armstrong explained one of the contributing factors to the decision was the ease of their transition to the Moon's weaker gravitational force. "My personal feeling was that the adaptation to 1/6 G was very rapid and was very pleasant, easy to work in, and I thought at the time that we were ready to go right ahead into the surface work and recommended that."[10] Armstrong continued:

> We wanted to do the EVA as soon as possible. We had thought even before launch that *if* everything went perfectly and we were able to touch down precisely on time, *if* we didn't have any systems problems to concern us, *if* we found that we could adapt to the one-sixth G lunar environment readily—then it

[9] "Greetings from Bruce McCandless II," Colin Mackellar, accessed March 10, 2021, Video, 5:07, https://vimeo.com/10151489.
[10] NASA, "Apollo 11 Technical Crew Debriefing," 10-7.

would make sense to go ahead and complete the EVA while we were still wide awake and not try to put that activity in the middle of the sleep period. In all candor, we didn't think that an early EVA was a very high probability. But as it turned out, we did land on time; there were no environmental or systems complications, so we chose to request permission to go ahead.[11]

After the shift change in Mission Control, Gene Kranz and Doug Ward prepared to walk over to the nearby Teague Auditorium in Building 1 for the post-landing press conference. Before they left the Control Center, they briefly talked about potential issues with the news media. Ward thought they needed to be prepared to discuss the communications dropouts during the powered descent. Kranz agreed and invited Don Puddy to come along. "I don't think Kranz, Puddy, or I had much to say during the walk from Mission Control to the news center," Ward recalls. "We were still processing the lunar landing, which seemed almost unreal to me."[12]

Steve Bales also attended the meeting. Leaving Mission Control, he noticed a peaceful group of demonstrators opposed to the Moon landing. Previously he was so focused on the mission he did not notice the protestors. Bales looked curiously at their sandals and long hair. Murray and Cox said, "It is unlikely they noticed Steve Bales as he hurried by, with his white shirt, his neatly tied tie, his NASA badge, pens in his pocket, square as could be."[13]

Milt Windler's Maroon Team settled into position in the Mission Control Center. Windler selected Maroon because it was the color of his alma mater, Virginia Tech.[14] Their shift lasted one hour before they handed off to Cliff Charlesworth's Green Team, which trained for the moonwalk.

Mike Collins searched for *Eagle* on the surface through *Columbia's* 28-power sextant. Knowing the exact coordinates would help them compute the next day's rendezvous data. Collins' "solo book" hanging

---

[11] Armstrong, Collins, Aldrin, *First on the Moon,* 306.
[12] Doug Ward, email message to author, April 4, 2019.
[13] Murray and Cox, *Apollo*, 355.
[14] Discussion with author at the Apollo 11 50th Anniversary Luncheon, Johnson Space Center, July 16, 2019.

around his neck not only had details of the rendezvous scenarios but also contained a LAM-2 Chart (Lunar Landing Area Map for Landing Site 2).[15] The chart was in a map package prepared for NASA by the US Army Topographic Command.[16]

The LAM-2 was a 1:100,000-scale chart of the landing site with a superimposed grid. The grid lines were labeled on the left edge with letters and the bottom edge with numbers. The controllers gave Collins specific search coordinates, which he wrote on a panel near *Columbia's* sextant. He hoped to see light reflecting off the LM but was unable.

Searching for *Eagle* forced Collins to maneuver the CSM into an attitude nearing "Gimbal Lock," where the inner and outer of the three gimbals around the guidance platform were in the same plane. If that happened, the platform would tumble out of control. Owen Garriott was the Maroon CAPCOM for the Command Module. After taking over, his first transmission to Mike Collins warned about this issue.

> GARRIOTT: We noticed you maneuvering very close to gimbal lock. I suggest you move back away. Over.
> COLLINS (JOKING): How about sending me a fourth gimbal for Christmas?[17]

HOLY COMMUNION

Sixty miles below, Armstrong and Aldrin had a chance to catch their breath. Aldrin retrieved his personal preference kit where he stored small packages containing wine and a wafer. He also packed a chalice and a small card containing words from the Gospel of John, chapter 15. "As Jesus said, I am the vine, you are the branches. Whoever remains in Me, and I in him, will bear much fruit; for you can do nothing without Me."

---

[15] NASA's Apollo Site Selection Board had named five possible sites in February of 1968 and ultimately chose Landing Site 2 for Apollo 11.

[16] The package contains one Lunar Landing Area Map at 1:100,000 scale, three Lunar Landing Site Maps at 1:25,000 scale, and 92 Lunar Surface Exploration Maps at 1:5,000 scale, covering the landing site with increasing detail. The same charts with white geology symbols are also included. A complete set of these charts for each mission can be found in the Lunar and Planetary Institute, Houston, Texas.

[17] A fourth gimbal would have solved the issue of gimbal lock. Gemini spacecraft had a four-gimbal system, but the extra weight caused designers to use only three for Apollo.

Not one to do things spontaneously, Aldrin also wrote his introduction on the card, which he read to Garriott and the listening world. "This is the LM pilot. I would like to take this opportunity to ask every person listening in, whoever and wherever they may be, to pause for a moment and contemplate the events of the past few hours and to give thanks in his or her own way." He read the John 15 verse silently and took Holy Communion while Armstrong watched respectfully.

Aldrin, an Elder at Webster Presbyterian Church south of Houston, described the scene to *Guideposts Magazine*. "I poured the wine into the chalice our church had given me. In the one-sixth gravity of the Moon, the wine curled slowly and gracefully up the side of the cup. It was interesting to think that the very first liquid ever poured on the Moon, and the first food eaten there, were communion elements."[18]

Later Armstrong and Aldrin ate dinner, then meticulously prepared for their EVA. They were more concerned with doing everything right than maintaining a schedule. The crew went through a pre-EVA checklist and donned their helmets, gloves, and Portable Life Support System (PLSS) backpacks. They hooked up their oxygen and cooling water, tested their communications, and prepared the cabin for depressurization.

There was a lot to do preparing for the moonwalk, and it took longer than expected. In their home near Houston, Janet Armstrong joked, "It's taking them so long because Neil is trying to decide about the first words he's going to say when he steps on the Moon. Decisions, decisions, decisions!"[19]

While the astronauts took their time, technicians at the Parkes radio telescope in Australia scrambled.[20] They planned for their 210-foot dish to receive television signals from the Moon later in the day, but the early EVA would begin with the Moon barely coming over the horizon. It was too low for the signals to be received on their main beam.

---

[18] "Guideposts Classics: When Buzz Aldrin Took Communion on The Moon," *Guideposts*, accessed November 9, 2020, https://www.guideposts.org/better-living/life-advice/finding-life-purpose/guideposts-classics-when-buzz-aldrin-took-communion-on-the-moon.
[19] Armstrong, Collins, Aldrin, *First on the Moon,* 309.
[20] Parkes, like most radio telescopes, receives signals but cannot transmit to spacecraft. Tracking stations, in comparison, do both.

The weather caused another major problem. While it was a hot, humid summer day in Houston, southeastern Australia faced frigid winter conditions. The wind limit for safely operating the dish was 30 miles per hour. Seventy mph gusts were hammering Parkes.[21]

Neil Mason was in the Parkes control room directly underneath the dish, trying to drive the enormous receiver down toward the horizon where the Moon was rising. He tells of their predicament:

> A gust of wind hit us, alarm bells sounded, and everyone was going, "What's that?" You're sitting there with a 1,200-ton dish rolling around the top of your head and there are creaks and bangs. The motors are trying to drive the dish down, the wind is pushing it back, and you sort of get a bit nervous. Rumbles were going on. If we obeyed the rules, we would just break off the track, stow the dish, and put it up on jacks.[22] We looked at each other and said, "What do we do now? What's going to happen?" The astronauts were just about to open the hatch, so we ignored the wind restrictions and waited for the Moon to rise. I was relieved when the boss said to keep going, as it took the decision off my shoulders.[23]

The Honeysuckle Creek tracking station was located 150 miles southeast of Parkes, near Australia's capitol city Canberra. Its 85-foot dish received telemetry signals from the Lunar Module, but the television camera was not yet activated. It was housed inside a Modularized Equipment Stowage Assembly (MESA) on the Lunar Module's descent stage. One of the outside walls of the LM was designed to fold down and provide a workbench and stowage area. Neil Armstrong would lower the MESA before descending the ladder.

While Parkes desperately hoped the Moon would rise high enough to get a video signal before the first step, television was the least of the concerns for Honeysuckle's Deputy Director Mike Dinn, who was monitoring the astronauts' biomedical telemetry. "I wasn't worried about the TV at all," recalls Dinn, "I was worried about the biomed. We were receiving intermittent signals from the Moon and did not know

---

[21] Lindsay, "Apollo 11 Essay," honeysucklecreek.net.
[22] This means pointing the dish straight up and putting jacks underneath to take stress off the gears and motors. Thanks to Colin Mackellar for the explanation!
[23] Lindsay, "Apollo 11 Essay," honeysucklecreek.net.

where the problem was. Television was not part of fulfilling Kennedy's charge, but returning the crew home safely was."[24]

The incoming signals contained the astronauts' vital signs and critical readings of the PLSS backpacks. "My operational judgment was that if we didn't have television, that in no way caused the mission to be a failure, but if there was a biomed issue, that really was a problem," said Dinn. "I'm still of that view."[25]

Operations Supervisor John Saxon agreed, adding, "It was so important we had a separate data line just for the biomed."[26] Fortunately, the telemetry soon stabilized.

Inside *Eagle*, Buzz Aldrin changed film in the 16-mm data acquisition camera, set the film rate to one frame per second, and mounted it pointing downward. The resulting field of view included the ladder and the area of the lunar surface where much of the EVA would take place. The 16-mm camera filmed the first steps, the planting of the flag, and other significant events during the moonwalk in case the television camera did not work.

As the crew waited for the cabin to finish depressurizing, Janet Armstrong said, "Make up your mind what you're going to say." Their son Rick said, "Knowing Dad, you can't tell. But it will be something good, though."[27]

The cramped cabin left little room for movement with both men fully suited. After *Eagle* depressurized, it was time to open the hatch and head outside. Armstrong joked, "Now comes the gymnastics." The hatch opened at 9:40 p.m. Houston time. It was centrally located with its bottom edge level with the floor and opened inwardly with its hinges on Aldrin's side. Armstrong got on his hands and knees and carefully backed through the 32-inch square opening with Aldrin directing him through the tight fit. Meanwhile, Mrs. Armstrong said, "He's still trying to think of what to say."[28]

Once outside the LM and on the porch, Armstrong pulled a D-Ring, lowering the MESA containing a Westinghouse television camera.

[24] Mike Dinn, Australia Apollo 11 Reunion, July 21, 2020.
[25] Mike Dinn, Australia Apollo 11 Reunion, July 21, 2020.
[26] Mike Dinn and John Saxon, Australia Apollo 11 Reunion, July 21, 2020.
[27] Armstrong, Collins, Aldrin, *First on the Moon,* 317.
[28] Armstrong, Collins, Aldrin, *First on the Moon,* 318.

Aldrin pushed in the circuit breaker, activating the camera.[29] The men experienced many problems with the it during training, so Armstrong did not expect it to work on the Moon. CAPCOM Bruce McCandless said they had a TV image, and Armstrong could not believe it:

> No one was more surprised than I. In all the preflight tests in which both the little image orthicon camera and I were simultaneously involved, I had never seen a picture! And that was when the camera and the television receiver were in the same building! So you can see why I was genuinely amazed when an actual picture was received in Houston via the very circuitous, complex, multi-frequency route from Mare Tranquilitatis via Australia.[30]

Two tracking stations received the video—Honeysuckle and Goldstone. Goldstone's version was displayed on the screen at the front of Mission Control, and McCandless described the image: "There's a great deal of contrast in it, and currently it's upside-down on our monitor, but we can make out a fair amount of detail."

The television camera had to be mounted upside down in the MESA, but later in the moonwalk would be set on a tripod right side up. Colin Mackellar explains, "Because of the camera handle and cabling, the only way to mount it would be upside down—the only flat surface on the camera was the top!"[31] The tracking stations were equipped with a large switch to invert the image for the early part of the EVA. The switch at Goldstone was initially in the wrong position.

Goldstone's 210-foot dish received good quality audio and video but had problems with their scan converter, which translated the images to television standards. The Moon still needed to be higher for Parkes to get a good signal on their main beam. It was time for Honeysuckle Creek to save the day.

Honeysuckle's 85-foot dish received a clear image from *Eagle*, and their inverter switch was in the correct position. As Neil Armstrong made his way down the ladder, technicians in Houston decided to use

---

[29] The spacecraft used push/pull circuit breakers. To activate equipment, the circuit breaker was pushed in, completing the circuit. The breaker was pulled to deactivate.
[30] Neil Armstrong, email message to Colin Mackellar, June 10, 2010.
[31] Colin Mackellar, email message to author, September 10, 2020.

the video from Honeysuckle and audio from Goldstone for the historic moment.

The camera lens poked through a hole in a protective blanket covering the tools in the MESA. It was mounted on an angle bracket which tilted the TV picture and caused the ladder to look much steeper than it actually was.[32]

Each LM leg had a footpad three feet in diameter and six inches deep.[33] Armstrong stood on the footpad beneath the ladder, then jumped up to see how difficult it would be to get back up at the end of the moonwalk. He returned to the footpad and described the lunar surface: "I'm at the foot of the ladder. The LM footpads are only depressed in the surface about one or two inches, although the surface appears to be very, very fine-grained as you get close to it. It's almost like a powder. The groundmass is very fine. Okay, I'm going to step off the LM now."

Artist Paul Calle recreated Armstrong's historic first step in a 4x8 foot oil painting, "The Great Moment." Michael Collins, himself an artist, said, "This painting of my friend Neil Armstrong by my friend Paul Calle combines for me the best of two worlds. NASA's technological achievements and an artist's exquisite interpretations of it." "The Great Moment," by Paul Calle, copyright Paul Calle, Chris Calle, CalleSpaceArt, www.callespaceart.com, chris@calleart.com.[34]

---

[32] Colin Mackellar, email message to author, October 11, 2022.
[33] NASA, "An Analysis and a Historical Review of the Apollo Program Lunar Module Touchdown Dynamics," 7.
[34] Special thanks to Chris Calle for permission to use "The Great Moment." To view more of Paul and Chris Calle's artwork, please go to www.callespaceart.com, and *see* Appendix D, pages 343–344.

While he held the ladder with his right hand, Armstrong's left boot touched lunar soil at 9:56 p.m. Houston time. "That's one small step for a man, one giant leap for mankind."

That was his intention. Audio does not clearly depict the indefinite article "a" before "man," and much has been written about the statement. Armstrong told biographer James Hansen, "I would hope that history would grant me leeway for dropping the syllable and understand that it was certainly intended, even if it wasn't said—although it might actually have been."[35]

In Mission Control, Apollo 12 Commander Pete Conrad sat beside off-duty FDO Jerry Bostick on a ledge behind the first-row console. Conrad looked at Bostick and asked, "What did he say?" Bostick replied, "Something about a giant leap for mankind." Conrad said, "That's just like Armstrong to say something profound."[36]

Armstrong continued, "The surface is fine and powdery. I can kick it up loosely with my toe. It does adhere in fine layers, like powdered charcoal, to the sole and sides of my boots. I only go in a small fraction of an inch, maybe an eighth of an inch, but I can see the footprints of my boots and the treads in the fine, sandy particles."

Armstrong's descriptive words had a specific purpose, informing the scientists about the lunar soil mechanics. Geologists trained him and Aldrin to report characteristics like adhesion and cohesion. The dust stuck to anything it touched, such as the astronauts' boots and spacesuits. Armstrong evaluated cohesion by how much the dust retained its shape after being compressed. He said, "I noticed in the soft spots where we have footprints nearly an inch deep that the soil is very cohesive, and it will retain a slope of probably 70 degrees along the side of the footprints."

He performed some preliminary tasks, including a panoramic photo sequence of the area and collecting a contingency sample of lunar rocks and soil in case an emergency caused them to leave suddenly. Armstrong was supposed to gather the contingency sample first but decided to take the photos while his eyes were still adapted to darkness in the Lunar

---

[35] Hansen, *First Man*, 268.
[36] Bostick, *The Kid from Golden*, 130–131. When Conrad jumped down from the ladder to the footpad during the Apollo 12 mission, he said, "Man, that may have been a small one for Neil, but that's a long one for me."

Module's shadow. His decision caused angst in the Control Center. Scientists cared more about the rocks and soil samples than pictures. However, Armstrong was the world's only subject-matter expert on moonwalking. He was comfortable in the lunar environment and confident he and Aldrin would perform the full EVA. He took care of the contingency sample after moving into the harsh sunlight away from the LM to avoid soil contaminated by the descent engine.

After the contingency sample, Armstrong guided Aldrin through the hatch and down the ladder while photographing his companion.

ALDRIN: Beautiful view!
ARMSTRONG: Isn't that something? Magnificent sight out here.
ALDRIN: Magnificent desolation.

They read a plaque attached to the leg of the LM., "Here men from planet Earth first stepped foot upon the Moon, July 1969 AD. We came in peace for all mankind." That was not the original version submitted by NASA to the White House. Nixon speechwriter William Safire said, "The first extraterrestrial problem to face Earthling speechwriters was in the editing of the copy of the plaque that was going to mark man's first landing on the Moon."[37] The original version had "first landed on the Moon," but senior speechwriter Pat Buchanan pointed out that "set foot" would be better since unmanned vehicles had already accomplished soft landings. Safire changed the tense of the original "we come in peace" to "we came in peace" so it would seem to be directed to future visitors from Earth instead of to present Moon inhabitants.[38]

Next, Aldrin freed the cable while Armstrong took the tripod and television camera about 50 feet northwest of the *Eagle*. He manually rotated the camera to give a panoramic view of the area. McCandless guided him for a final orientation with *Eagle* in the picture.

---

[37] William Safire, *Before the Fall: An Inside View of the Pre-Watergate White House,* (New York: Doubleday), 1975, 143–144.
[38] Safire, *Before the Fall,* 144.

THE PROUDEST DAY OF OUR LIVES

After deploying the American flag, Armstrong prepared to collect a bulk sample of lunar soil when they received a special request from Houston.

> MCCANDLESS: We'd like to get both of you in the field of view of the camera for a minute. Neil and Buzz, the President of the United States is in his office now and would like to say a few words to you, over.
> ARMSTRONG: That would be an honor.
> PRESIDENT NIXON: Hello, Neil and Buzz. I'm talking to you by telephone from the Oval Room at the White House, and this certainly has to be the most historic telephone call ever made from the White House.[39] I just can't tell you how proud we all are of what you have done. For every American, this has to be the proudest day of our lives. And for people all over the world, I am sure that they, too, join with Americans in recognizing what an immense feat this is. Because of what you have done, the heavens have become a part of man's world. And as you talk to us from the Sea of Tranquility, it inspires us to redouble our efforts to bring peace and tranquility to Earth. For one priceless moment in the whole history of man, all the people on this Earth are truly one; one in their pride in what you have done and one in our prayers that you will return safely to Earth.
> ARMSTRONG: Thank you, Mr. President. It's a great honor and privilege for us to be here representing not only the United States, but men of peace of all nations, men with interests and a curiosity, and men with a vision for the future. It's an honor for us to be able to participate here today.
> PRESIDENT NIXON: And thank you very much, and all of us look forward to seeing you on the *Hornet* on Thursday.
> ALDRIN: We look forward to that very much, sir.

Nixon's words, "All the people on this Earth are truly one," accurately described the experience of most people tuning in to the

---

[39] The words "from the White House" were not heard on the television audio due to the echo from the Moon. Colin Mackellar points out that for the 2019 documentary *Apollo 11*, Archive Producer Stephen Slater insisted on using audio recorded from the White House containing these words so it does not sound like such a grandiose claim. Australia Apollo 11 Reunion, July 21, 2020.

broadcast. An estimated 650 million people watched or listened worldwide.[40] Local agencies provided screens and television sets in public locations all over the planet. Thousands gathered in New York City's Central Park. On the other side of the globe, the RSL Memorial Theatre in Carnarvon, Australia, placed a small TV set on a table for the local townspeople. It was the first time many of them had ever watched television.[41]

Geologists in Houston did not appreciate Nixon's phone call. Gerry Griffin assisted Flight Director Cliff Charlesworth and listened on the communication loop. "The scientists were going ape, thinking the President was wasting valuable time on the surface," said Griffin. He thought, "Here is the President of the United States talking to the astronauts, and the scientists are yelling, 'knock it off—we've got to get back gathering up stuff.'" Griffin shook his head and grinned.[42]

The July 24, 1969 edition of Carnarvon's *The Northern Times* describes how the citizens stayed up all night listening to the landing and spent the morning viewing the historic moonwalk:

> Many viewed the small TV screen set on the theatre's stage through binoculars, and one of the town's crack riflemen had his telescope focused on the astronauts as they walked on the Moon surface, collecting samples, planting the American flag, and carrying out various scientific experiments. Anyone who saw the event on the TV set or heard the radio coverage of Apollo 11 must have been thrilled and proud to know that Carnarvon, too, was playing its part.[43]

After the phone call, the astronauts were behind their timeline and hurried through checklists sewn onto their cuffs. Armstrong collected the bulk sample, using a scoop to fill one of the rock boxes with soil. He and Aldrin set up a Passive Seismology Experiment to detect

---

[40] NASA, "Apollo 11 Overview,"
https://www.nasa.gov/mission_pages/apollo/missions/apollo11.html.
[41] Lindsay, "Apollo 11 Essay," honeysucklecreek.net.
[42] Gerry Griffin, interview by author, July 13, 2020.
[43] For a fascinating look at the efforts made to provide satellite TV coverage to that region, *see* Colin Mackellar's "How the moonwalk was seen live in Western Australia," accessed October 30, 2019,
https://www.honeysucklecreek.net/Apollo_11/A11_TV_Perth.html.

moonquakes, a Laser Reflector to measure the precise distance from the Earth to the Moon using laser ranging, and a Solar Wind Experiment to study the composition of electrically charged particles emitted by the Sun.

One hour and forty-five minutes into the moonwalk, McCandless alerted them that only 10 minutes remained. Aldrin collected two core tube samples measuring the stratification of material beneath the surface. As he hammered the tubes, he met serious resistance below the surface. Aldrin informed Houston, "I hope you're watching how hard I have to hit this into the ground, to the tune of about five inches."

While Aldrin was preparing the core tubes, Armstrong took the opportunity to explore an area of scientific interest. He scampered back to a 33-meter diameter crater he flew over in the final minute before landing, now called Little West Crater. It lies 60 meters behind *Eagle*, and Armstrong hoped it was deep enough to expose some lunar bedrock. When he arrived, he realized it had not pierced through to the bedrock, but he collected some samples and took a photographic sequence of the area.

Dr. Lee Silver, who trained the Apollo astronauts in geology, was impressed Armstrong thought to do this:

> I have to tell you that what Neil did in the shortest period of time that anybody had was so brilliant from this point of view of providing the materials to the scientists, that nobody can claim to have exceeded it in production per minute. He was really outstanding. And I had nothing to do with it; it doesn't reflect on my work at all. And we didn't even begin to realize what he had done. He broke rules. He had a very strict protocol which said, "You will never leave the field of view of the camera." Neil Armstrong recognized that just beyond the field of the camera was a rim of craters covered with rocks and dust, which had been excavated from a little deeper than everywhere else. He had a very special box for bringing back good samples with a special seal on it, and for about seven or eight minutes, you couldn't see Neil. The focus was on the second man, Buzz Aldrin. What was Neil doing? He stuffed that box so fast and so full of lunar samples, and that has yielded so much material of value to science that I don't think

anybody has, at the rate that Neil did, achieved efficiency. And that was in eight or ten minutes. He's a very special guy.[44]

Aldrin went up the ladder first at the end of the EVA. He entered the hatch, then used the Lunar Equipment Conveyer (LEC) to transfer the rock boxes and film magazine from Armstrong on the surface into the cabin. The LEC was similar to a clothesline attached to a pulley. Aldrin then tossed down a memorial packet with an Apollo 1 patch honoring Grissom, White, and Chaffee. It also contained medals honoring deceased cosmonauts Yuri Gagarin and Vladimir Komarov, a microminiaturized photo print of goodwill letters from world leaders, and a gold olive branch. Armstrong placed the packet near the footpad. After two hours and thirteen minutes on the lunar surface, he climbed the ladder to join Aldrin.

> McCANDLESS: Neil, this is Houston. Did you get the Hasselblad magazine?
> ARMSTRONG (JOYFULLY): Yes, I did. And we got about, I'd say, 20 pounds of carefully selected, if not documented, samples.[45]
> McCANDLESS (LAUGHING): Roger. Well done.

Deke Slayton came on the air-to-ground loop for a chat with the crew.

> SLAYTON: I just want to let you guys know that since you're an hour and a half over your timeline and we're all taking a day off tomorrow, we're going to leave you. See you later.
> ARMSTRONG: I don't blame you a bit.
> SLAYTON: That's a real great day, guys. I really enjoyed it.
> ARMSTRONG: Thank you. You couldn't have enjoyed it as much as we did.
> ALDRIN: It was great.
> SLAYTON: Sure wish you'd hurry up and get that trash out of there, though.
> ARMSTRONG: We're just about to do it.

---

[44] Leon T. Silver, interview by Carol Butler for NASA Johnson Space Center Oral History Project, May 5, 2002, 10–11.

[45] The flight plan called for documented samples, meaning rocks and soil to be photographed on the surface, identified, and stored in individual bags. Armstrong and Aldrin ran out of time and did not document or separate the samples.

Keeping their promise to the boss, Armstrong and Aldrin discarded their PLSS backpacks, trash, and other unnecessary items out of the hatch to save weight for their ascent the next day.

> **McCANDLESS:** We observed your equipment jettison on the TV, and the passive seismic experiment recorded shocks when each PLSS hit the surface. Over.
> **ARMSTRONG:** You can't get away with anything anymore, can you?

At the Honeysuckle Creek tracking station, video technician Ed von Renouard was operating the slow-scan console as well as the scan converter, which took the TV signals from the Moon and changed it to US television standards.[46] He recorded portions of the moonwalk by pointing his personal Super 8 camera at the monitor so successfully that his film was often better than what the world saw on television. When he heard Slayton's comment about discarding the trash, he activated his camera again and filmed the scenes still being transmitted to the tracking stations.[47] He captured the backpacks tumbling down from the hatch; it is the only known complete recording of the event. Neil Armstrong signed off at 3:30 a.m., over 21 hours after the historic day began.

Buzz Aldrin read John 15:5 from the top of this card before taking Communion on the Moon. He also copied Psalm 8:3-4 below it, which he read on the trip back to Earth. Photo credit: LM Otero/AP Photo.

---

[46] Colin Mackellar, "The Apollo 11 Television Broadcast," accessed November 1, 2019, https://honeysucklecreek.net/Apollo_11/index.html.
[47] Colin Mackellar, email message to author, October 13, 2022.

Buzz Aldrin examines the landing gear and footpad.
Photo credit: NASA AS11-40-5902.

Aldrin deploys the passive seismic experiment.
Photo credit: NASA AS11-40-5948.

Armstrong flew over this crater just before landing,
and sampled it near the end of the moonwalk.
It is now known as Little West Crater.
Photo credit: NASA AS11-40-5954.

Neil Armstrong in Eagle after the moonwalk.
Photo credit: NASA AS11-37-5528.

21

*I AM LIKE
A NERVOUS BRIDE*

ON THEIR WAY HOME

At 9:30 a.m. Monday, CAPCOM Ron Evans gave a wake-up call to
Mike Collins in *Columbia*:

> **EVANS:** *Columbia, Columbia*; good morning from Houston.
> **COLLINS:** Morning, Ron.
> **EVANS:** Hey, Mike, how's it going this morning?
> **COLLINS:** I don't know yet. How's it going with you?

Ten minutes remained in Collins' frontside pass. Before they lost
signal, Evans let him know about a change of plans for the next
revolution, the last full rev before lunar liftoff. They wanted him to
perform a landmark tracking exercise using the small crater known as
"130 prime" near the landing site, which John Young used on Apollo
10. At the same time, Armstrong and Aldrin would have their
rendezvous radar tracking *Columbia* as it passed overhead.

> **EVANS:** *Columbia*, this 130 is the little bitty crater, John
> Young's crater, that you tracked prior to descent. We want
> this for one last fix on your plane.
> **COLLINS:** Fine. Understand. Thank you.[1]

---

[1] Landmark 130 is on the southwest rim of Sabine D, a small crater located about 20
km north-northeast of the Apollo 11 landing site. Sabine D was later named "Collins"

Then Collins went around the corner for the 23rd time.

*PUBLIC AFFAIRS OFFICER (TERRY WHITE): During this pass and
the short conversation toward the end of the frontside pass
with Columbia, the network transmitters have been arranged
so that the transmissions would not disturb the crew of
Eagle, who at this time should be asleep. Not since Adam has
any human known such solitude as Mike Collins is
experiencing during the 47 minutes of each lunar revolution
when he's behind the Moon with no one to talk to except his
tape recorder aboard Columbia. While he waits for his
comrades to soar with Eagle from Tranquility Base and
rejoin him for the trip back to Earth, Collins, with the help of
flight controllers here in Mission Control Center, has kept
the Command Module's system going "pocketa-pocketa-
pocketa." At 120 hours, 12 minutes ground elapsed time, this
is Apollo Control.*

Two critical questions lingered in Mission Control. First, where
was the Lunar Module? From Armstrong's descriptions during the
descent, they were a few miles downrange (west) of the target. Pete
Williams, an expert in the Real-Time Computer Complex (RTCC) on
the first floor underneath the MOCR, suggested they use the LM
rendezvous radar to track the Command Module flying overhead. Once
they had that data, they could run the rendezvous program in the RTCC
backward to determine the relative position of *Eagle* compared to
Columbia.[2] FDO David Reed liked the idea, took his headset off, and
spoke to Flight Director Milt Windler face to face.[3] Windler agreed, and
Ron Evans called the procedure up to Aldrin.

Eagle's rendezvous radar tracked *Columbia* as it passed overhead
for the final time before liftoff, thus giving controllers the data they
needed to compute a liftoff time with the most fuel-efficient rendezvous.
David Reed explains, "Rendezvous is not dependent on absolute
position (i.e., latitude and longitude) but rather where is one relative to

---

by the International Astronomical Union (IAU). Its diameter of 3.2 km (2 miles)
makes it a challenging target for amateur astronomers, requiring the right lighting,
good seeing, and great optics.
[2] Jerry Bostick, interview by author, August 25, 2020.
[3] David Reed, "Looking Back," in *From the Trench of Mission Control to the Craters
of the Moon*, 192–194.

the other. Think of it like a quarterback passing to a receiver. He throws not looking at what yard line he or his receiver is on, but by mentally calculating/knowing their relative positions."[4]

There was one more question to be answered. What caused the computer overload during the descent? Steve Bales was still exhausted from the previous day when he began his ascent shift. He watched the beginning of the moonwalk from the flight controller lounge. After the astronauts planted the flag, Bales went to the bunk room across the hall and slept until it was time for the ascent team to report for duty.[5]

He hoped someone would have resolved the problem overnight. MIT Professor David Mindell writes, "Moments after the touchdown, the phone at the MIT Instrumentations Laboratory [IL] began ringing like a 1202 Program Alarm. NASA was calling and wanted to know what went wrong, demanding an explanation and a fix before the LM lifted off the surface in a few hours."[6]

The MIT team frantically worked on the problem throughout the night. One of their engineers, George Silver, finally solved it. Since an abort was a realistic scenario during the descent, there was a desire to have the rendezvous radar turned on so it would be ready in an emergency. Both the radar and guidance computer ran on AC (Alternating Current) power. They were connected to the same power supply when tested on the ground. But in the Lunar Module during flight, they were connected to different power supplies. The voltage for the radar was out of phase with the voltage for the computer. When the computer incorporated data from the radar during the descent, the issue stole 15 percent of the computer's processing ability.[7]

Program 63, which ran the first part of the descent, used 85 percent of the computer's capacity. The added 15 percent from the voltage issue put the computer right at the limit. Any additional requirement, such as monitoring the Delta-H or Program 64 used during the Approach Phase, caused an executive overflow. The computer bailed out, set off an alarm, rebooted instantaneously, and carried on with only the most essential tasks.

---

[4] David Reed, email message to author, January 10, 2020.
[5] Steve Bales, email message to author, January 20, 2020.
[6] David Mindell, *Digital Apollo*, 227.
[7] David Mindell, *Digital Apollo*, 228.

The rendezvous radar's mode switch had three positions. Two were manual options, and the third was an automatic setting called "LGC" because it put the Lunar Module Guidance Computer in control of the radar. Silver noticed the phasing issue before and knew it caused problems when the radar was set to one of the manual modes. When he learned the switch was in a manual mode during the descent, he immediately got in touch with his representatives in Houston.

Word got to Jack Garman in the Mission Control back room 20 minutes before lunar liftoff. Garman needed to tell Bales the mode switch for the rendezvous radar must be set to LGC. He also wanted the radar circuit breakers to be pulled, deactivating the radar until it was needed.

Steve Bales writes, "No flight controller likes last-minute procedure changes and I was no exception—in fact, I may have been worse than most."[8] While Mission Control and the entire planet sweated bullets over the possibility of two astronauts being stranded on the Moon, Garman called from the support room. Leading up to critical phases of a mission, the Guidance Officer had an assistant, call sign "Yaw," whom Garman normally would have told.[9] This time, he asked to speak directly to Bales:

> **GARMAN:** George Silver thinks he understands the problem, and it sounds very realistic for the descent.
> **BALES:** It's a little late. What can we do?
> **GARMAN:** We want to change one switch position to guarantee we don't get hit again.
> **BALES:** What's that?
> **GARMAN:** We want the rendezvous radar circuit breakers pulled as they are now, but we want the mode switch in LGC position.
> **BALES:** That's what it's in, I believe, Jack. I'll check.
> **GARMAN:** We want that verified. We only have 21 minutes.

Bales called the Control Officer, Hal Loden, inquiring about the switch setting. Loden informed him the rendezvous radar mode switch

---

[8] Steve Bales, "From Iowa to the Trench—A Guido's Tale," in *From the Trench of Mission Control to the Craters of the Moon*, 255.

[9] Steve Bales, email message to author, January 20, 2020.

was currently in SLEW, one of the manual settings. Bales demonstrated his complete trust in Garman, immediately replying, "That's a problem. We're going to get bit if we don't get it in LGC." Bales quickly called the Flight Director for the ascent, Glynn Lunney, to campaign for the change.

> **BALES:** Flight, Guidance.
> **LUNNEY:** Go.
> **BALES:** The back room has found a breakthrough. They NEED the rendezvous radar mode switch to LGC. They still want to keep the rendezvous radar circuit breaker pulled, but they want to make sure the rendezvous radar mode switch is to LGC. They found something in the software/hardware interface. I don't know what it is yet, but they really need this.
> **LUNNEY:** Are we positive of that?
> **BALES:** They are as positive as can be.
> **LUNNEY:** Is there some kind of reason?
> **BALES:** We'll try to get a better explanation, but I recommend we get that set up. It is an MIT recommendation.

Lunney, showing implicit trust in Bales and his back room, called CAPCOM Ron Evans.

> **LUNNEY:** We want to leave the rendezvous radar circuit breakers pulled, but we want the rendezvous radar mode switch in LGC.
> **EVANS** (HESITATING): Okay... uh... I hope they got it squared away there.
> **LUNNEY:** So do I.

Evans was uncomfortable changing a carefully planned procedure so close to such a critical event. After more hesitation, he asked Lunney, "What do they say is the effect of that?"

Steve Bales was listening on the same loop and inserted himself into the conversation. "Flight, Guidance. We could very well run the risk of putting the computer cycle much higher than it should be. This is the latest recommendation; I recommend we do it."

Bales later recalled, "We did not have time to discuss the details with the MIT team in Boston; it was either believe them or not. After getting assurance this couldn't make things worse, and knowing enough about the system to have some confidence it wouldn't, I was ready to make the change."[10]

One minute and forty seconds after Bales first notified Lunney, the final decision was made. Lunney instructed Evans, "At this time, we go with it, CAPCOM."

While Evans relayed the change to Armstrong, someone from MIT called Bales through his loop with a more technical explanation. Steve did not have time to hear the whole statement. He quickly replied, "We're going to do it. We'll take your word for it." By "your" word, Bales meant MIT in general. He did not recognize the voice on the other end of the conversation until listening to the tape later. It was George Silver.[11]

The crew needed one more piece of information. Lunney asked Bales and Loden whether they liked the Primary Guidance and Navigation System (PGNS, pronounced "pings") or the secondary Abort Guidance System (AGS, pronounced "aggs") for the rendezvous. Without much thought, Bales replied, "PGNS"

That quick decision gives insight into the world of the Mission Techniques group under Bill Tindall's leadership. "If something like the alarms had happened during a previous mission, Bill would have conducted many sessions determining whether we should select PGNS or AGS for ascent. Normally we selected PGNS unless there was good reason not to do so. But in this case, couldn't one argue there might be reason to pick AGS?"[12]

However, the primary system performed all of its significant functions well during the descent despite the program alarms, and MIT thought they understood why those alarms occurred. Plus, the secondary AGS could be used in case there was an issue with the primary "although rendezvous with the AGS would require a lot more manual intervention by the crew. And you really needed a good reason to recommend AGS because this would have probably set the crew on edge," explained Bales. "So, in sum, I really didn't worry too much about recommending

[10] Steve Bales, "From Iowa To the Trench—A Guido's Tale," in *From the Trench of Mission Control to the Craters of the Moon*, 255.
[11] Steve Bales, email message to author, January 20, 2020.
[12] Steve Bales, email message to author, January 20, 2020.

PGNS, although if we had stayed on the surface two days, no doubt there would have been many hours discussing the subject. Sometimes not having days to ponder things works out well." [13] Loden also recommended the primary system.

> LUNNEY: CAPCOM, our recommendation for launch is PGNS. We're cleared for takeoff.

Evans repeated those words to the crew. Aldrin replied, "Roger, understand. We're number one on the runway." There were 17 minutes left before liftoff.

Jim Lovell once again sat nervously beside Charlie Duke. This time, they were not at the CAPCOM console but on a step at the rear of the MOCR. Fourteen months earlier, Duke recommended NASA switch to the Rocketdyne injector for the LM ascent engine. Now the wisdom of that decision would be tested before the watching world. It simply had to work.

John Houbolt, the chief proponent for the Lunar Orbit Rendezvous (LOR) mode, was confident the engine would fire. "That was one of the best-tested engines in the universe." [14] The simple engine only had a few moving parts and was either on or off, with no adjustable throttle. Like the LM descent engine and Service Module SPS engine, it used a hypergolic propellant combination which ignited the moment the fuel came in contact with the oxidizer. The resulting flame was colorless.

The big moment approached. "I am like a nervous bride," Collins admits in his autobiography, "I have never sweated out any flight like I am sweating out the LM now. My secret terror for the last six months has been leaving them on the Moon and returning to Earth alone; now I am within minutes of finding out the truth of the matter." [15]

Onboard *Eagle*, Armstrong said to Aldrin, "At five seconds before launch, I'm going to get Abort Stage and Engine Arm." Aldrin counted down to ignition. "Nine, eight, seven, six, five. Abort Stage, Engine Arm, Ascent, Proceed." The engine worked perfectly, and its exhaust knocked over the flag and blew pieces of Kapton foil from the descent

---

[13] Steve Bales, email message to author, January 20, 2020.
[14] Armstrong, Collins, Aldrin, *First on the Moon*, 354.
[15] Collins, *Carrying the Fire*, 411–412.

stage in all directions. Aldrin reported, "We're off. Look at that stuff go all over the place. Look at that shadow. Beautiful." *Eagle* gained altitude and headed west toward the lunar terminator, passing over stunning craters in lengthening shadows, a dramatic sight.

> **ALDRIN:** There's Ritter out there. There it is, right there; there's Schmidt. Man, that's impressive looking, isn't it?
> **EVANS:** *Eagle*, Houston. You're looking good.
> **ARMSTRONG:** Looking good here. It's a spectacular ride.

Seven minutes after liftoff, the ascent engine shut down right on time, much to Collins' relief. "One little hiccup and they are dead men," wrote Collins, who was holding his breath the entire time of the ascent engine burn.[16]

Over the next few hours, *Eagle* performed maneuvers with its thrusters to gradually intersect with *Columbia*. As he passed over the Sea of Tranquility during the next revolution, Collins radioed Armstrong and Aldrin. "*Eagle, Columbia*. Passing over the landing site, it sure is great to look down there and not see you."

An hour later, Collins could view *Eagle* in his sextant. "Well, I see you don't have any landing gear."

Armstrong answered, "You're not confused on which end to dock with, are you?"

Onboard *Eagle*, Armstrong told Aldrin, "One of those two bright spots is bound to be Mike."

"How about picking the closest one?"

"Good idea," said the Commander.

As *Eagle* approached, Collins called it, "The best sight of my life: Neil and Buzz returning!"[17] He used the sequence camera for video photography while taking still shots with the Hasselblad. Then Collins said to Armstrong and Aldrin, "I've got the Earth coming up already. It's fantastic!" Andrew Chaikin describes the well-known image of the LM with Earth rising over the lunar horizon: "All of humanity captured in a single photograph, minus only himself, the photographer."[18]

---

[16] Collins, *Carrying the Fire*, 412.
[17] Collins, *Carrying the Fire*, caption of photo after page 358.
[18] Chaikin, *A Man on the Moon*, 225.

Armstrong held *Eagle* steady as Collins successfully performed the docking maneuver. Then Collins removed the hatch, plus the docking probe and drogue from the tunnel. They joyfully greeted one another. Collins recalls, "We cavort about a little bit, all smiles and giggles over our success, and then it's back to work as usual, Neil and Buzz prepare the LM for its final journey and I help them transfer equipment into *Columbia*."[19]

During the process, Armstrong showed Collins a zipped beta cloth pouch containing the sealed contingency sample bag. "If you want to have a look at what the Moon looks like, you can open that up and look. Don't open the bag, though." Collins unzipped the pouch and looked at the Teflon bag with the lunar soil.[20] He was surprised at the dark color of the dust.

ARMSTRONG (LAUGHING): You'd never have guessed, huh?
COLLINS: What was that bag?
ARMSTRONG: Contingency sample.
COLLINS: Rock?
ARMSTRONG: Yes, there are some rocks in it too. You can feel them, but you can't see them. They're covered with that graphite. It looks like powdered graphite to me.

They transferred the priceless rock boxes, film, and other necessary items to *Columbia*. The crew vacuumed and cleaned, removing as much lunar dust as possible, and then it was time to jettison *Eagle*. "The separation was slow and majestic," said Armstrong. "We were able to follow it visually for a long time."[21]

The next item of business was the burn to get them home, the Trans-Earth Injection (TEI). Charlie Duke was CAPCOM for that phase of the mission. Collins told him, "It's a happy home up here. It'd be nice to have some company. As a matter of fact, it'd be nice to have a couple of hundred million Americans up here."

Duke said, "Roger. Well, they were with you in spirit anyway, at least that many. We heard that yesterday after you made your landing,

---

[19] Collins, *Carrying the Fire*, 417.
[20] NASA, "Apollo 11 Preliminary Science Report," 117.
[21] NASA, "Apollo 11 Technical Crew Debriefing," 13-1.

the *New York Times* came out with the largest headlines they've ever used in the history of the newspaper."

"Save us a copy."

Armstrong said, "I'm glad to hear it was fit to print."

Gene Kranz's White Team reviewed data and evaluated the spacecraft systems. Near the end of the 30th frontside pass, Duke radioed the crew, "You are Go for TEI."

The burn would occur on the Moon's far side, ten minutes after Apollo 11 emerged into daylight. Until then, the crew's meticulous final preparations would happen in the dark. They eased the tension with humor.

Collins said, "You know, if you hit this hand controller like you do in the simulator—minimum impulse, just bang it—it'll bang over and bang back, and it'll fire two opposing pulses, and you get nothing."

Instrumentation told them they were in the right attitude, but they waited until they broke out of darkness for visual confirmation. Firing the engine in the wrong direction would spoil the day. Their Gemini flights came into mind, especially the retrofire burns which brought their spacecraft out of orbit and toward the ocean. Gemini crews carefully checked their alignment before retrofire burns, and now the Apollo 11 astronauts were doing the same. Ten minutes before ignition, they observed their final lunar sunrise.

COLLINS: Aha. I see a horizon.
ARMSTRONG (LAUGHING): It looks like we are going *forward*. It is most important that we be going *forward*. There's only one really bad mistake you can make here.
COLLINS (LAUGHING): Shades of Gemini retrofire. Let's see, the motors point this way and the gases escape that way, therefore imparting a thrust that-a-way.

They quickly set their minds back on business. Their next visual check would come two minutes before ignition when the lunar horizon was supposed to be at a predetermined location out their window.

ARMSTRONG: Two minutes to get our horizon check at 10 degrees.

COLLINS: Yes, sneaking up on there. Looks pretty darn good. Looks like we're darn near there.
ARMSTRONG: Just about midnight in Houston town.
COLLINS: Okay, coming up on two minutes, and this horizon check is going to be—would you believe—perfect?
ALDRIN: I hope so.
COLLINS: Fantastic. First time we ever got a perfect horizon check. Spent too many hours in the simulator looking for an unreal horizon. Alright, horizon check passes.

The Trans-Earth Injection burn was flawless. Mike Collins could not contain his appreciation for the Service Module's Service Propulsion System: "Beautiful burn. SPS, I love you! You are a jewel!" They were on their way home.

There was a buzz in Mission Control as they counted down the time to acquisition of signal if the burn was nominal. Gene Kranz told the flight controllers, "Hey everybody, let's keep the chatter down. We've got about a minute to acquisition."

PAO (DOUG WARD): This is Apollo Control. We're now less than one minute from reacquiring Apollo 11. When last we heard from the spacecraft, all systems were looking very good and we were in very good shape for the Trans-Earth Injection maneuver. At this point, Apollo 11 should be about 10,000 pounds lighter and on its way back to Earth. We're now 32 seconds from reacquiring. We'll stand by for communications with the spacecraft.

Onboard Apollo 11, Armstrong asked if anyone had a choice greeting for Houston when they acquired communication. Inside Mission Control, Charlie Duke watched his screen anxiously, then raised his arms triumphantly as the data appeared right on schedule. It was six minutes after midnight.

DUKE: Hello Apollo 11. Houston. How did it go? Over.
ARMSTRONG: Time to open up the LRL doors, Charlie.
DUKE: Roger. We got you coming home. It's well stocked.
DUKE: And Apollo 11, Houston. All your systems look real good to us. We'll keep you posted.
ARMSTRONG: Rog.

COLLINS: Hey, Charlie boy, looking good here. That was a beautiful burn. They don't come any finer.

PAO (WARD): That Trans-Earth Injection burn was very close to nominal. At this time, we show the spacecraft traveling at a speed of 7,603 feet per second. The velocity already beginning to drop off. And we're at an altitude of 445 nautical miles now from the Moon.

Apollo 11 rapidly departed, photographing the Moon as its size visibly decreased out the window. At 1:30 a.m., Deke Slayton, who was CAPCOM for Alan Shepard's historic Mercury flight, got on the air-to-ground loop with a message for his crew:

This is the original CAPCOM. Congratulations on an outstanding job. You guys have really put on a great show up there. I think it's about time you powered down and got a little rest, however. You've had a mighty long day here. Hope you're all going to get a good sleep on the way back. I look forward to seeing you when you get back here. Don't fraternize with any of those bugs in route, except for the *Hornet*.[22]

Armstrong replied, "Thank you, boss. We're looking forward to a little rest and a restful trip back. And see you when we get there."

"Rog. You've earned it."

Charlie Duke returned with some information about *Eagle's* Lunar Module Guidance Computer, which Mission Control had been monitoring to see how long it could last without being cooled. "Hello, Apollo 11. For your information, the LGC in *Eagle* just went belly-up at seven hours."

Armstrong replied, "Very good. That was the death of a real winner."

At 2:20 a.m., it was time for the White Team to hand off to Milt Windler's Maroon Team. Duke informed them, "11, Houston. Shift change time here. White Team bids you good night. We'll see you tomorrow. Over."

---

[22] The *USS Hornet* was the recovery aircraft carrier. The crew would quarantine for three weeks to make sure no harmful life forms were brought back from the Moon.

They thanked one another for the great work done in space and in Mission Control, then Owen Garriott took over as CAPCOM for the overnight shift.

Collins welcomed him, "Good morning, Owen. You purple people keep funny hours."

Apollo Guidance Computer Staff Support Room (SSR).
Jack Garman is in the dark jacket, second from left.
Photo credit: NASA S-69-34209, Jack Garman, and Colin Mackellar.

Honeysuckle Creek tracking station near Canberra, Australia.
Photo credit: Hamish Lindsay, Colin Mackellar.

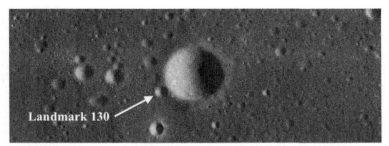

Lunar landmark 130 is the small crater on the southwest rim of Sabine D (Collins). John Young and Mike Collins used 130 for landmark tracking exercises. See page 218 for more context.
Photo credit: NASA, Lunar Orbiter 5, 5074.

"The best sight of my life." Eagle returns from the lunar surface to the great delight of Michael Collins. Photo credit: NASA AS11-44-6642.

22

THE SPACECRAFT GAVE A LITTLE JUMP AS IT WENT THROUGH THE SPHERE

RETURN TO EARTH, 12:40 p.m. TUESDAY, JULY 22

The Apollo 11 crew woke up from a well-deserved 10-hour rest period. For six months, the crew endured intense pressure. "The three of us were just the tip of a gigantic technological iceberg," said Collins. "We felt the weight of the world on our shoulders. Everyone was looking; we were worried we were going to screw something up."[1]

With the burden lifted, they could relax and have some fun. Flight Director Cliff Charlesworth's Green Team was on duty, and Bruce McCandless was CAPCOM as the spacecraft approached a landmark on the journey home. Public Affairs Officer Doug Ward alerted the media the spacecraft was about to cross the imaginary line into Earth's sphere of influence. CAPCOM informed the crew:

MCCANDLESS: You're leaving the lunar sphere of influence.
COLLINS: Roger. Is Phil Shaffer down there?

---

[1] CBSnews.com, "NASA Legends Remember the Nerve-Wracking Moments Before Apollo 11 Landing," July 15, 2019, https://www.cbsnews.com/news/apollo-11-moon-landing-anniversary-nasa-legends-remember-the-nerve-wracking-moments/.

McCANDLESS: Negative. But we've got a highly qualified team on in his stead.

COLLINS: Rog. I wanted to hear him explain it again at the press conference.

McCANDLESS: Okay.

COLLINS: That's an old Apollo 8 joke but tell him the spacecraft gave a little jump as it went through the sphere.

McCANDLESS (LAUGHING): Okay. I'll pass it on to him.

CHARLESWORTH (SARCASTICALLY TO McCANDLESS): Tell him, *"Thanks a lot!"*

McCANDLESS (LAUGHING, TO COLLINS): Thanks a lot, and Dave Reed is sort of burying his head in his arms right now.

COLLINS (LAUGHING): Roger that.

Reed was thinking, "It was gonna be a fun item to rub Phil with at the splashdown parties!"[2]

Meanwhile, Neil Armstrong carried on the friendly banter. "Those guys down there in the trench did a pretty good job this flight."[3] McCandless agreed.

Armstrong continued, "We don't want them to give up yet, though."

Charlesworth told McCandless on the Flight Director loop, "That will never happen.

CAPCOM relayed it to the crew. "No. They'll hang in there for about another 47 hours or so."

At 3:00 p.m., Apollo 11 performed the first midcourse correction on their trip home. The crew used their thrusters for a 10-second retrograde burn, scrubbing five feet per second from their velocity. An hour later, it was time for questions about the landing site. The geologists in the science room thought they had pinpointed the location but wanted to make sure.

McCandless asked, "For 64 thousand dollars, we're still trying to work out the location of your landing site, Tranquility Base.[4] We think

---

[2] David Reed, email message to author, January 30, 2020.

[3] The "Trench" refers to the first row in Mission Control, where the trajectory specialists operated.

[4] "The $64,000 Question" was a television show in America during the 1950s. Contestants were asked general knowledge questions. The prize for correct answers doubled as the questions became more difficult, and the final question was worth $64,000.

it is located on LAM-2 chart at Juliet 0.5 and 7.8. Do you still have those charts on board?

> COLLINS: Stand by. They're packed.
>
> McCANDLESS: You may not have to unpack it. The position which I just gave you is slightly west of West Crater. I guess it's about two-tenths of a kilometer west of it, and we were wondering if Neil or Buzz had observed any additional landmarks during descent, lunar stay, or ascent which would confirm or disprove this. One thing that we're wondering about is that if you were at this position, you would have seen the Cat's Paw during ascent just up to the north of your track. Over.

Armstrong replied, "We were looking for the Cat's Paw, too, thinking we were probably downrange. But I think that might have been West Crater that we went across in landing."

McCandless asked one more question. "Did you observe any small craters with conspicuously blocky rims?"

Armstrong added a detail Mission Control did not know:

> Well, aside from the real big one that we went over, I guess there were none in our area. I took a stroll back to a crater behind us that was maybe 70 or 80 feet in diameter and 15 or 20 feet deep and took some pictures of it. It had rocks in the bottom of a pretty good size, considerably bigger than any that were out on the surface. But we, apparently at 15 feet or so, had not gotten below the regolith.[5] We were essentially showing no bedrock, at least in the walls of the crater at that depth. Over.[6]

That settled it. Geologist Gene Shoemaker later said, "Had Neil told us about the small crater behind the LM, we could have pinpointed them right then between twelve to twenty meters."[7]

Shortly after 6:00 p.m., the White Team came back on duty. One hundred fifty thousand miles above, the astronauts were listening to

---

[5] Regolith is the surface layer of loose rock and soil covering the lunar bedrock.
[6] The small crater explored by Armstrong is now known as "Little West Crater."
[7] Armstrong, Collins, Aldrin, *First on the Moon,* 388.

music on their small Sony tape recorder. Armstrong's selections included "The New World Symphony" by Dvorak. He also brought Les Baxter's "Music Out of the Moon" and Jonathan King's "Everyone's Gone to the Moon," which became a favorite of Mike Collins.

While Charlie Duke was settling in as CAPCOM, the Apollo 11 crew indulged in a little mischief. At the end of Armstrong's tape was what Collins described as "a jangling cacophony of bells, whistles, shrieks, and unidentifiable sounds."[8] They held the screeching tape player beside the microphone and pushed the transmit button. All of a sudden, the air-to-ground communication loop was filled with a loud racket. The crew laughed in the background.

Flight controllers had no idea what was going on. They looked at each other and asked, "Can you tell what that was?"

Duke checked with the crew. "Apollo 11, Houston. You sure you don't have anybody else in there with you?

Mike Collins acted like nothing happened and said, "Houston, Apollo 11. Say again, please."

"We had some strange noises coming down on the downlink, and it sounded like you had some friends up there."

"Where did the White Team go during their off hours, anyway?"

Technicians in Mission Control tried to figure it out. George Egan, a Philco-Ford employee, got on the Network loop and said, "There were some horrible, terrible noises. I've never heard anything like that before, kind of like an Indian attack."

Network controller Richard Stachurski laughed and asked his co-worker Chester Brantley to check with Goldstone to see if they heard any weird noises coming from the spacecraft on the downlink.

The contact at Goldstone said, "I've been awaiting your call; I sure did."

Stachurski and Brantley then reported to Charlesworth and Duke it came from the spacecraft and not from any technical issues on the ground.[9]

---

[8] Collins, *Carrying the Fire*, 427.
[9] Stachurski and Brantley were friends and neighbors in the outside world, and experienced the highs and lows of Apollo together. Stachurski and his wife were shopping on Friday, January 27, 1967. When they came home, Brantley knocked on the door and informed them of the tragic fire claiming the lives of Grissom, White, and Chaffey. Richard Stachurski, email message to author, February 3, 2020.

PUBLIC AFFAIRS OFFICER (TERRY WHITE): This is Apollo Control. Still no explanation on the weird noises emanating from Apollo 11, if indeed it is from Apollo 11, and it's reported from Network that it's being received on the downlink at two different stations in the Manned Space Flight Network. Perhaps it will all shake out later in the mission as to what these strange noises are.[10]

Soon after 8:00 p.m., the crew turned on their television camera. The signals went from *Columbia* through the Goldstone tracking station into Mission Control, where they were displayed on a large screen by an Eidophor projection system. Since the Eidophor was enlarging analog video signals, the resulting image was fuzzy.

ARMSTRONG: Are you picking up our TV signals?
DUKE: That's affirmative. We have it up on the Eidophor now. The focus is a little bit out. We see Earth in the center of the screen. And we see some land masses in the center, at least I guess that's what it is. It's very hazy at this time on our Eidophor. Over.
ALDRIN: I believe that's where we just came from.
DUKE: It is, huh? Well, I'm really looking at a bad screen here. Stand by one.

Charlie Duke turned around to see the television set behind his CAPCOM console which had a clearer picture, then looked at the big screen again to compare the two images. He had to admit, "Hey you're right!"

Duke immediately caught grief from the flight control team. His boss Deke Slayton could not resist jumping on the air-to-ground loop.

SLAYTON: It's not bad enough, not finding the right landing spot; you haven't even got the right planet!
DUKE: I'll never live that one down.

---

[10] It remained a mystery until the astronauts' post-quarantine press conference when the crew was asked about the noises. They grinned at each other. Armstrong said, "We are guilty. We sent the whistles and bells [laughing]. We had our little tape recorder which we used to record our comments during the flight in addition to playing music in the lonely hours, and we thought that we'd just share some of that with the people in the Control Center." Armstrong, Collins, Aldrin, *First on the Moon,* 511, fn. 7.

COLLINS: We're making it get smaller and smaller here to make sure that it really is the one we're leaving.
DUKE: All right. That's enough, you guys.

While Duke continued to catch flack inside Mission Control, Collins set up the camera for views inside the Command Module, starting with the Commander by the rock boxes. Armstrong said,

> We know there are a lot of scientists from a number of countries standing by to see the lunar samples, and we thought you'd be interested in seeing that they really are here. These two boxes are the Sample Return Containers. They are vacuum-packed containers that were closed in a vacuum on the lunar surface, sealed, and then brought inside the LM and put inside these fiberglass bags, zippered and re-sealed around the outside and placed in these receptacles in the side of the Command Module. These are the two boxes, and as soon as we get onto the ship, I'm sure these boxes will immediately be transferred, and then delivery started to the Lunar Receiving Laboratory. These boxes include the samples of the various types of rock, the groundmass of the soil, the sand and silt, the particle collector for the solar wind experiment, and the core tubes that took depth samples of the lunar surface.

Buzz Aldrin demonstrated ham spread on bread and gave a physics lesson about spinning gyroscopes. Mike Collins showed how to drink water out of a spoon and shot water in his mouth from the water gun, explaining, "It's sort of messy. I haven't been at this very long. It's sort of the same system that the Spaniards used to drink out of wineskins at bullfights, only I think this is even more fun. We'll be seeing you, kids."

> DUKE: Thank you from all us kids in the world, here in the MOCR, who can't tell the Earth from the Moon.
> COLLINS: Roger. Stand by one, and we'll get you that Earth one.
> DUKE: Looks like you need a wineskin up there, Mike.
> COLLINS: That'd be nice.

Armstrong had the camera and showed Earth through the window. He asked, "You have a picture now, Houston?"

DUKE: That's affirmative. I refuse to bite on this one, though. You tell us.

ARMSTRONG: Okay. This should be getting larger, and if it is, it's the place we're coming home to. No matter where you travel, it's always nice to get home.

DUKE: We concur, 11. We'll be happy to have you back.

ARMSTRONG: This is Apollo 11, signing off.

Things quieted down as the crew approached their rest period. Then Aldrin started exercising around 10:30 p.m. which was recorded on his biomedical sensors. His heart rate reached 150 beats per minute. "Buzz, you brought the surgeon right out of his chair," Duke said.

After a few quiet minutes, Collins said to Duke, "The ol' White Team has really got a busy one tonight, huh?"

"Oh, boy. We're really booming along here with all this activity. Can barely believe it."

"What are you doing? Sitting around with your feet up on the console drinking coffee?"

Duke laughed and said, "You must have your X-ray eyes up. You sure can see a long way."

"Yeah. We're watching you as well, you know."

"This is the highlight of the day—Buzz's exercise for the surgeon—the highest heartbeat ever seen in manned spaceflight. The surgeon is about to die!"

Then Neil Armstrong played one of his selections from "Music Out of the Moon." It was recorded in 1947 using a theremin which produced eerie futuristic sounds. The music was loud enough to trigger the voice-operated exchange transmitter switch (VOX), so Mission Control had to listen. When it mercifully stopped, Duke quickly said, "Thank you, 11. We appreciate you turning that off."

ARMSTRONG: Charlie, could you copy our music down there?

DUKE: Rog. We sure did. We're wondering who made your selections.

ARMSTRONG: That's an old favorite of mine. It's an album made about 20 years ago called *Music Out of the Moon*.[11]
DUKE: Roger. It sounded a little scratchy to us, Neil. Either that or your tape was a little slow.
ARMSTRONG: It's supposed to sound that way.
COLLINS: It sounds a little scratchy to us too, but the czar likes it.
DUKE: That's what we figured. He and his 40,000 votes.
DUKE: For your info, it looks like you are about 150,000 miles out now.
COLLINS: It's getting appreciably larger now. Its looking more like the world.
DUKE: I'll never hear the last of that one about that Earth/Moon business during the TV.
COLLINS: You'll have fun at the press conference after this shift then, won't you?
DUKE: It's 2:30. Everybody will be asleep. I'm going to sneak off through the back way.

WEDNESDAY, JULY 23, 1969, 11:30 a.m.

CAPCOM (OWEN GARRIOTT): Apollo 11, Houston. Are you up and at 'em yet?
ARMSTRONG: Well, we're up, at least, Owen.

Twenty-four hours remained before splashdown. After some housekeeping duties, it was time for Garriott to report the latest news, including a story about a newborn.

In Memphis, Tennessee, a young lady who is presently tipping the scales at eight pounds, two ounces, was named Module by her parents, Mr. and Mrs. Eddie Lee McGhee. "It wasn't my idea," said Mrs. McGhee. "It was my husband's." She said she had balked at the name Lunar Module McGhee because it didn't sound too good, but apparently they have compromised on just Module.[12]

---

[11] Janet Armstrong listened to the squawk box in their home and was surprised to hear the song. When her husband returned, she said, "Where on Earth did you find that? We used to listen to it all the time." Armstrong, Collins, Aldrin, *First on the Moon*, 390.
[12] Module McGhee grew up to be a mother and special education teacher who loves her unique name. *See* Seth Borenstein, "Named for Lunar Lander, Module McGhee

A few hours later, it was time for a shift change in Houston, and the Green team took over. Bruce McCandless was CAPCOM, and Collins asked him, "Did Dave Reed get to explain the lunar sphere at the press conference?

"No, but your comments about Phil Shaffer and the explanations were quoted in the paper last night."

"Uh oh."

McCandless asked, "Do you want to say anything more while you're on the line?"

"He's right. He's absolutely right."

At 6:00 p.m., the crew began their final television broadcast, giving their impressions of the previous eight days.

ARMSTRONG: Good evening. This is the Commander of Apollo 11. A hundred years ago, Jules Verne wrote a book about a voyage to the Moon. His spaceship, *Columbia*, took off from Florida and landed in the Pacific Ocean after completing a trip to the Moon. It seems appropriate to us to share with you some of the reflections of the crew as the modern-day *Columbia* completes its rendezvous with the planet Earth and the same Pacific Ocean tomorrow. First, Mike Collins.

COLLINS: This trip of ours to the Moon may have looked to you simple or easy. I'd like to assure you that has not been the case. The Saturn V rocket which put us into orbit is an incredibly complicated piece of machinery, every piece of which worked flawlessly. This computer up above my head has a 38,000-word vocabulary, each word of which has been very carefully chosen to be of the utmost value to us, the crew. This switch which I have in my hand now has over 300 counterparts in the Command Module alone, this one single switch design. In addition to that, there are myriads of circuit breakers, levers, rods, and other associated controls. The SPS engine, our large rocket engine on the aft end of our Service Module, must have performed flawlessly, or we would have been stranded in lunar orbit. The parachutes up above my

Rockets On," July 21, 2009, *The San Diego Union-Tribune*, https://www.sandiegouniontribune.com/sdut-us-apollo-name-072109-2009jul21-story.html.

head must work perfectly tomorrow, or we will plummet into the ocean. We have always had confidence that all this equipment will work and work properly, and we continue to have confidence that it will do so for the remainder of the flight. All this is possible only through the blood, sweat, and tears of a number of people. First, the American workmen who put these pieces of machinery together in the factory. Second, the painstaking work done by the various test teams during the assembly and the re-test after assembly. And finally, the people at the Manned Spacecraft Center, in management, in mission planning, in flight control, and last but not least, in crew training. This operation is somewhat like the periscope of a submarine. All you see is the three of us, but beneath the surface are thousands and thousands of others, and to all those, I would like to say thank you very much.

ALDRIN: Good evening. I'd like to discuss with you a few of the more symbolic aspects of the flight of our mission, Apollo 11. As we've been discussing the events that have taken place in the past two or three days here onboard our spacecraft, we've come to the conclusion that this has been far more than three men on a voyage to the Moon. More, still, than the efforts of a government and industry team. More, even, than the efforts of one nation. We feel that this stands as a symbol of the insatiable curiosity of all mankind to explore the unknown. Neil's statement the other day upon first setting foot on the surface of the Moon, "This is a small step for a man, but a great leap for mankind," I believe sums up these feelings very nicely. We accepted the challenge of going to the Moon; the acceptance of this challenge was inevitable. The relative ease with which we carried out our mission, I believe, is a tribute to the timeliness of that acceptance. Today, I feel we're fully capable of accepting expanded roles in the exploration of space. In retrospect, we have all been particularly pleased with the call signs that we very laboriously chose for our spacecraft, *Columbia* and *Eagle*. We've been particularly pleased with the emblem of our flight, depicting the US eagle bringing the universal symbol of peace from the Earth, from the planet Earth to the Moon, that symbol being the olive branch. It was our overall crew choice to deposit a replica of this symbol on the Moon.

Personally, in reflecting on the events of the past several days, a verse from Psalms comes to mind. "When I consider Thy heavens, the work of Thy fingers, the Moon and the stars, which Thou hast ordained; what is man, that Thou art mindful of him?"

ARMSTRONG: The responsibility for this flight lies first with history and with the giants of science who have preceded this effort. Next, to the American people, who have through their will, indicated their desire. Next, to four administrations and their Congresses for implementing that will. And then, to the agency and industry teams that built our spacecraft, the Saturn, the *Columbia*, the *Eagle*, and the little EMU; the spacesuit and backpack that was our small spacecraft out on the lunar surface. We'd like to give a special thanks to all those Americans who built those spacecraft, who did the construction, design, the tests, and put their hearts and all their abilities into those craft. To those people, tonight, we give a special thank you, and to all the other people that are listening and watching tonight, God bless you. Good night from Apollo 11.

PAO (DOUG WARD): That brief view of the Earth came from 91,371 nautical miles out in space after a brief, and sincere, and moving transmission from the Apollo 11 spacecraft. This is Apollo Control at 177 hours, 45 minutes.

Charlie Duke took over as CAPCOM after the television broadcast. He informed them about bad weather in the planned recovery site. "The weather is clobbering in at our targeted landing point due to scattered thunderstorms. We don't want to tangle with one of those, so we're going to move your aim point downrange." The Apollo Command Module had an offset center of gravity which created lift during entry, just like the Gemini capsule.[13] A rolling action would steer it toward the landing area.

It was time for Apollo 11 to sign off for the night. Duke said, "It's a good night from the White Team for the last time. We'll be off when

---

[13] Usually the lift was aimed up or down. Aiming the lift vector up would lengthen the glide patch, while pointing it down would bite into the atmosphere for more drag and a quicker descent.

you wake up in the morning. It's been a pleasure working with you guys. It was a beautiful show from all three of you. We appreciate it very much, and we'll see you when you get out of the LRL."

> **ARMSTRONG:** Okay, Charlie. Thanks to you and all the White Team for a great job down there all the way through.
> **ALDRIN:** Thank you very much, Charlie. Thanks.
> **DUKE:** Thanks to you guys, too. And Mike, you get your chance at landing tomorrow. No go-arounds.
> **COLLINS:** Rog. You're going to let me land closer to Hawaii, too, aren't you?
> **DUKE:** That's right, sir.

As that conversation occurred, Duke suffered through one last dig from the bunch in Mission Control. Public Affairs Officer Terry White commented on the scene:

> *This is Apollo Control. All goodnights having been said, the crew of Apollo 11 is now preparing to get their 10 hours rest in their last night in space. Here in the Control Center, on one of the 10 by 10 Eidophor television projectors, a drawing has been projected on the screen ribbing CAPCOM Charlie Duke for his slight error yesterday on the television pass where he mistook the Moon for Earth. It has the spacecraft midway between the Moon and Earth, and it says, "Neil, I just spotted a continent on the Moon."*
> *"Charlie, the camera is on the Earth now."*

Flight Dynamics Officer David Reed during Apollo 11. Screen capture from NASA 16mm film, provided by Stephen Slater and Apollo 11 in Real Time.

23

THE BEST SHIP TO COME DOWN THE LINE. GOD BLESS HER

CAPTAIN FRANK BROWN

Qantas Airline pilot Frank Brown was a huge space enthusiast and realized his flight QF596 from Sydney to Honolulu would fly parallel to the Apollo 11 entry ground track 600 miles away. However, the flight schedule would put them in prime position a few hours before the historic event. Upon inquiry, NASA recommended they delay the flight by three hours and nineteen minutes to have the best view.[1] Qantas agreed and advertised the flight as a "Lunar Special."[2]

Qantas offered those traveling in First Class "An Apollo 11 Out of This World" souvenir menu including Sea of Tranquility Oysters en Capsule, Duckling a l'Armstrong au Champagne, Roast Loin of Lamb Aldrin, Chilled Lunar Lobster with Collins Salad, and Flight Path Fruit Salad. All passengers received a commemorative certificate.[3]

---

[1] Helen Styles, "Boeing Saw Re-entry Spectacle," *Sydney Sun-Herald*, July 27, 1969. https://www.honeysucklecreek.net/images/Apollo_11/Sun-Herald_27-7-69.jpg.
[2] "$811.80 to See Apollo Blaze By," *The Age (Melbourne)*, July 23, 1969. https://www.newspapers.com.
[3] Qantas, "Onboard the Qantas Flight Which Was Out of This World," July 19, 2019. https://www.qantasnewsroom.com.au/roo-tales/onboard-the-qantas-flight-which-was-out-of-this-world.

Flying high above the Pacific Ocean, the Boeing 707 pilot searched the skies to the north. At about 2:30 a.m. local time, right on schedule, the fiery trail of the Command Module appeared 66 miles high as it streaked through the atmosphere. The bright spectacle was visible for over three minutes, lighting up the skies near Gilbert and Ellice Islands. Additionally, the abandoned Service Module burned up in the atmosphere in view of the 82 passengers and 13 flight crew members. Brown gave a running commentary to the passengers during the event.[4]

AFTER SPLASHDOWN

The flight of Apollo 11 ended in the Pacific Ocean, 900 miles southwest of Hawaii, on Thursday morning, July 24. Splashdown occurred at 11:50 a.m. in Houston and soon after local sunrise. Keeping with tradition, Mission Control did not begin their celebration until the crew arrived safely on the deck of the recovery aircraft carrier, the *USS Hornet*. PAO Jack Riley proudly stated, "The flags are waving, and the cigars are being lit up. And clear across the big board in front is President John F. Kennedy's message to Congress of May 1961: 'I believe that this nation should commit itself to achieving the goal, before this decade is out, of landing a man on the Moon and returning him safely to Earth.'" On a screen to the right was the Apollo 11 emblem. Above it was written, "Task Accomplished... July, 1969."

Hundreds of flight controllers and support staff packed into the Control Center, celebrating one of the greatest achievements in human history. David Reed had the foresight to secure the Flight Dynamics plot board used during the launch and had key individuals autograph it. Armstrong, Aldrin, and Collins would sign it after their time in quarantine. The plot board eventually ended up on display at the Smithsonian Institution.[5]

Much has been written of the splashdown party in Houston that night. However, the flight crew was not able to make it. They left the helicopter on the *Hornet* hangar deck wearing Biological Isolation

---

[4] "Bird's-Eye View of Apollo 'Fireball,'" *The Age (Melbourne)*, July 26, 1969. https://www.newspapers.com.
[5] David Reed, "Looking Back," *From the Trench of Mission Control to the Craters of the Moon*, 194.

Garments and went immediately into a NASA trailer called the Mobile Quarantine Facility. After a preliminary medical exam and quick showers, they spoke with President Nixon as he stood outside the trailer.

Looking at the grinning astronauts behind a window and speaking through an intercom system, Nixon began, "Neil, Buzz, and Mike, I want you to know that I think I'm the luckiest man in the world. And I say this not only because I have the honor to be President of the United States, but particularly because I have the privilege of speaking for so many in welcoming you back to Earth." He invited them to a State Dinner in Los Angeles on August 13, three days after completion of the quarantine period.

In his remarks, President Nixon mentioned Frank Borman, who was with him in the White House during important phases of the mission. "Frank Borman says you're a little younger by reason of having gone into space, is that right? Do you feel that way, a little younger?"

Collins replied, "We're *a lot* younger than Frank Borman!"

Nixon called Borman over to the conversation. "Come on over, Frank, so they can see you. Are you going to take that lying down?"

Collins said, "It looks like he has aged in the last couple of days."

Borman replied, "Mr. President, you know we have a poet in Mike Collins, and he really gave me a hard time for describing in words of 'fantastic' and 'beautiful.' And I counted them. In three minutes up there, you used four fantastics and two beautifuls."

The conversation then took on a weighty tone:

PRESIDENT NIXON: Well, just let me close off with this one thing: I was thinking, as you know, as you came down, and we knew it was a success, and it had only been eight days, just a week, a long week, that this is the greatest week in the history of the world since the Creation, because as a result of what happened in this week, the world is bigger, infinitely. And also, as I am going to find on this trip around the world, and as Secretary Rogers will find as he covers the other countries in Asia, as a result of what you have done, the world has never been closer together before. We just thank you for that. I only hope that all of us in government, all of us in America, that as a result of what you have done, we can do our job a little better. We can reach for the stars, just as you have reached so far for the stars. We don't want to hold you any longer. Does anybody have a last request? How about promotions? Do you think we could arrange something?

ARMSTRONG: We're just pleased to be back and very honored that you were so kind as to come out here and welcome us back, and we look forward to getting out of this quarantine and talking without having glass between us.

PRESIDENT NIXON: And incidentally, the speeches that you have to make at this dinner can be very short. And if you want to say fantastic or beautiful, that's all right with us. Don't try to think of any new adjectives; they've all been said. I think the millions that are seeing us on television now would feel as I do, that in a sense, our prayers have been answered. I think it would be very appropriate if Chaplain Piirto, the Chaplain of this ship, were to offer a prayer of thanksgiving. If he would step up now. Chaplain, thank you.

LT. COMDR. JOHN A. PIIRTO, USN CHAPLAIN OF THE *HORNET*: Let us pray. Lord, God, our Heavenly Father, our minds are staggered and our spirits exultant with the magnitude and precision of this entire Apollo 11 mission. We have spent the past week in communal anxiety and hope as our astronauts sped through the glories and dangers of the heavens. As we try to understand and analyze the scope of this achievement for human life, our reason is overwhelmed with abounding gratitude and joy, even as we realize the increasing challenges of the future. This magnificent event illustrates anew what man can accomplish when purpose is firm and intent corporate. A man on the Moon was promised in this decade. And, though some were unconvinced, the reality is with us this morning in the persons of astronauts Armstrong, Aldrin, and Collins. We applaud their splendid exploits and we pour out our thanksgiving for their safe return to us, to their families, to all mankind. From our inmost beings, we sing humble yet exuberant praise. May the great effort and commitment seen in this project Apollo, inspire our lives to move similarly in other areas of need. May we the people, by our enthusiasm and devotion and insight, move to new landings in brotherhood, human concern, and mutual respect. May our country, afire with inventive leadership and backed by a committed followership, blaze new trails into all areas of human cares. See our enthusiasm and bless our joy with dedicated purpose for the many needs at hand. Link us in friendship with peoples throughout the world as we strive together to better the human condition. Grant us peace, beginning in our own hearts, and a mind attuned with goodwill toward our neighbor. All this we pray as our thanksgiving rings out to Thee, in the name of our Lord. Amen.[6]

The Command Module *Columbia* was brought aboard *Hornet* and attached by a plastic tunnel to the astronauts' quarantine trailer so the rock boxes and film could be retrieved, passed through a decontamination transfer lock, and flown to the LRL in Houston.[7] Having access to the spacecraft which treated him so well, Mike Collins wanted to express his feelings. "On the second evening, I climb back on board its charred carcass, and on the wall of the lower equipment bay, just above the sextant mount, I write: 'Spacecraft 107—alias Apollo 11—alias *Columbia*. The best ship to come down the line. God bless her. Michael Collins, CMP.'"[8]

QUARANTINE

Hornet steamed to Hawaii, where the Mobile Quarantine Facility was loaded on a transport plane and flown to Houston. It landed during the early hours of Sunday, July 27. All of the inhabitants went into the Lunar Receiving Laboratory (LRL), "well stocked," as advertised to the crew by Charlie Duke during the mission. The LRL had private bedrooms for each astronaut, a dining area, a kitchen, a living room, a recreation area, and laboratories for the rock samples.

---

[6] The American Presidency Project, "Remarks to Apollo 11 Astronauts Aboard the *USS Hornet* Following Completion of Their Lunar Mission," accessed October 3, 2021. https://www.presidency.ucsb.edu/documents/remarks-apollo-11-astronauts-aboard-the-uss-hornet-following-completion-their-lunar.
[7] William Compton, *Where No Man Has Gone Before: A History of Apollo Lunar Exploration Missions*, (Washington, DC: National Aeronautics and Space Administration, 1989), 147.
[8] Collins, *Carrying the Fire*, 446. This note, plus the LAM-2 chart coordinates Collins wrote near the sextant as he searched for *Eagle* on the lunar surface, are still visible in *Columbia*. The Smithsonian Institution photographed the spacecraft's interior with a camera attached to a boom before the 50th anniversary of Apollo 11, and Collins' writing on the wall is clearly seen.

It also contained, by design, rodents that were exposed to the lunar samples for signs of pathogens. During the Apollo 11 50th anniversary celebrations, Mike Collins recalled, "We had a huge colony of white mice. The three of us had gone to the Moon. That was either an international triumph, or it was a total disaster depending on the health of the white mice that we had. If the mice lived, everything was fine."[9]

"Twelve men are in absolute quarantine here because of something that probably does not exist," reported Harold Schmeck of the *New York Times*. "The team includes technicians, research analysts and medical technologists, stewards, a photographer, a public information specialist, and the Manned Spacecraft Center's chief of clinical laboratories, Dr. Craig Fischer."[10]

The "public information specialist" Schmeck mentioned was NASA Public Affairs Officer John McLeaish, who did the commentary on the launch phase of the mission. McLeaish and the doctors, technicians, support staff, and mice began their quarantine before *Eagle* left the lunar surface. He held daily post-flight meetings with the press pool through a glass wall, updating them on the medical condition of the humans and mice, and giving summaries of the astronauts' technical debriefings.

While in quarantine, Armstrong asked McLeaish for advice on handling questions from the media about his first words on the Moon. Some in the press were making a big deal about whether it was one small step for "man" or "a man." Armstrong wanted McLeaish's counsel on how to respond. McLeaish replied, "You are the expert on it. Whatever you say is what history is going to record. If you say your words were 'One small step for a man,' that's what history will record." Armstrong's intent was to say "a man," and that is what he thought he said.[11]

Doug Ward expressed what most people felt. "To me, and I think to the world, they understood what Armstrong was getting at. That was

---

[9] James Rogers, "Apollo 11's Michael Collins Recounts the Crew's Three-Week Quarantine on Their Return from The Moon," July 22, 2019, https://www.foxnews.com/science/apollo-11-michael-collins-quarantine-moon.

[10] Harold M Schmeck, Jr, "Crew That Will Join Astronauts in Quarantine Already Living in Isolation Building," *New York Times*, July 23, 1969, https://timesmachine.nytimes.com.

[11] Doug Ward, interview by author, March 16, 2019.

the impression the line made to everyone who heard it."[12] Before they left quarantine, McLeaish took his copy of the Apollo 11 flight plan to Armstrong and asked if he would sign the page where he first stepped foot upon the Moon. Armstrong thought it was a great idea and signed it, "One small step for a man, one giant leap for mankind, Neil A. Armstrong." Ward adds, "As far as I know, that is the only such autograph in existence."[13]

AUGUST 12, 1969

By August 10, it was clear neither the astronauts nor the test mice had become sick from exposure to the lunar surface, so the quarantine was terminated. Two days later, representatives of the news media from the United States and around the world had their first opportunity to hear from the three heroes. At 10:00 a.m. on Tuesday, August 12, two hundred reporters were on hand for a televised press conference at the Manned Spacecraft Center in Houston. NASA's Assistant Administrator for Public Affairs, Julian Sheer, opened the proceedings by introducing the astronauts, who gave a 45-minute presentation on the mission. Afterward, there was a period of questions from the media. Some inquiries were technical, while others focused on the emotional and philosophical aspects of the flight.

One reporter asked if there was ever a moment during the moonwalk when Armstrong and Aldrin "were just a little bit spellbound by what was going on." Armstrong replied, "About two and a half hours."

Later, when asked about getting behind schedule in the opening minutes on the lunar surface, the Commander said, "We plead guilty to enjoying ourselves. As Buzz mentioned earlier, we are recommending that we start future EVAs with a 15 or 20-minute period to get these kinds of things out of the way and to get used to the surface."

A media representative mentioned the popular criticism the space program is "a misplaced item on a list of national priorities" and asked, "how do you view space exploration as a relative priority compared with

---

[12] Doug Ward, interview by author, March 16, 2019.
[13] Doug Ward, interview by author, March 16, 2019.

the present needs of the domestic society and the world community at large?"

Armstrong said, "Of course, we all recognize that the world is continually faced with a large number of varying kinds of problems and that it's our view that all those problems have to be faced simultaneously. It's not possible to neglect any of those areas, and we certainly don't feel that it's our place to neglect space exploration."

The crew was asked for their impressions of the meaning of the first landing on another celestial body.

ALDRIN: Well, I believe that what this country set out to do was something that was going to be done sooner or later whether we set a specific goal or not. I believe that from the early space flights, we demonstrated a potential to carry out this type of mission. And again, it was a question of time until this would be accomplished. I think the relative ease with which we were able to carry out our mission came after a very efficient and logical sequence of flights. This demonstrated that we were certainly on the right track when we took this commitment to go to the Moon. What this means is that many other problems, perhaps, can be solved in the same way by making a commitment to solve them in a long-time fashion. We were timely in accepting this mission of going to the Moon. It might be timely at this point to think in many other areas of other missions that could be accomplished.

COLLINS: To me, there are near and far-term aspects to it. In the near term, I think it a technical triumph for this country to have said what it was going to do a number of years ago and then, by golly, do it just like we said we were going to do—not just, perhaps, purely technical, but also a triumph for the nation's overall determination, will, economy, attention to detail, and a thousand and one other factors that went into it. That's short term. I think, long term, we find for the first time that man has the flexibility or the option of either walking this planet or some other planet, be it the Moon or Mars, or I don't know where. And I'm poorly equipped to evaluate where that may lead us.

ARMSTRONG: I just see it as a beginning—not just this flight, but this program, which has really been a very short piece of human history, an instant in history—it's a beginning of a new age.

AUGUST 13, 1969

The following day, August 13, was a coast-to-coast flurry of activity, with ticker tape parades in New York, Chicago, and Los Angeles. That evening President Richard Nixon hosted a State Dinner in honor of the three astronauts at the Century Plaza Hotel in Los Angeles. Present were the Governors of 44 states, including future President Ronald Reagan, then the Governor of California. Also on the guest list of 1,440 were NASA officials, 50 astronauts, Charles Lindbergh, General James Doolittle, Bob Hope, Jack Benny, Red Skelton, and Billy Graham.[14]

Governor Reagan began the evening by recounting the hectic day. "Our three guests of honor started the day under the ticker tape and confetti in New York where they stood and received the plaudits of the people of the Empire State, and then later in the day were in Chicago where again they stood beneath the ticker tape and confetti and watched the fireworks on the lake shore, and we hope in the days to come they will recall that in California we bid them sit down and we fed them and gave them to drink." He then led a toast to the President of the United States.

NASA Administrator Dr. Thomas Paine read a citation posthumously awarding NASA's highest decoration, the Distinguished Service Medal, to astronauts Gus Grissom, Ed White, and Roger Chaffey. Mrs. Betty Grissom and Mrs. Pat White were present to receive the medals earned by their husbands.

Dr. Paine then presented the Group Achievement Award to the Apollo 11 Mission Operations Team. "For exceptional service in planning and exemplary execution of mission operational responsibilities for Apollo 11, the first manned lunar landing mission. The distinguished performance of this team was decisive in the success of the first extraterrestrial exploration mission, a major milestone in the

---

[14] "Dinner Guest List Reflects, Past, Present, and Future Triumphs," *New York Times*, August 14, 1969, https://timesmachine.nytimes.com.

advancement of mankind." Steve Bales received the award for the team. President Nixon added, "This is the young man when the computers seemed to be confused, and he could have said "Stop," or when he could have said, "Wait," said "Go!"[15]

Astronauts Armstrong, Collins, and Aldrin each received the Presidential Medal of Freedom, the highest civilian honor presented to an American citizen. Collins spoke first:

> Mr. President, here stands one proud American, proud to be a member of the Apollo team, proud to be a citizen of the United States of America, proud to be an inhabitant of this most magnificent planet [provoking laughter from the audience]. As I looked at it from nearly a quarter of a million miles away three weeks ago, the people of New York, of Chicago, and of Los Angeles were far from my mind frankly [more laughter], but tonight they are very close to my mind. I wish that each and every one of you could have been with us today to see their enthusiasm and the magnificent greeting which they gave us upon our return. And, of course, now the Freedom Medal. I simply cannot express in words what that means to me, but I would like to say thank you very much.

Buzz Aldrin spoke next:

> It is an honor in a sense that goes to all Americans who believed, who persevered with us. Across this country today, we saw how deeply they believe. We saw it by their spontaneous enthusiasm and warmth of greeting for us. Our flight was your flight. We flew *Eagle* and *Columbia* with your hands helping us on the controls, and your spirit behind us. When Neil and I saluted the flag, all Americans, I think, saluted it with us…. Never before have travelers been so far removed from their homelands as we were, yet never before have travelers had so many human beings at their right hand. There are footprints on the Moon. Those footprints belong to each and every one of you, to all of mankind. And they are there because of the blood, the sweat, and the tears of

---

[15] "California Dinner for Apollo 11 Astronauts, part 1," Audio, 11:46, accessed January 15, 2020.
https://archive.org/details/CaliforniaDinnerForApollo11Astronauts/California+Dinner+for+Astronauts_8-13-1969_side+1_.wav.

millions of people. These footprints are a symbol of the true human spirit. Thank you, Mr. President."

An emotional Neil Armstrong spoke last:

We were privileged to leave on the Moon a plaque, endorsed by you, Mr. President, saying, "for all mankind." I was struck this morning in New York by a proudly-waved but uncarefully-scribbled sign. It said, "Through you, we touched the Moon." Through you, we touched the Moon. It was our privilege today to cross the country to touch America. I suspect that perhaps the most warm, genuine feeling that all of us could receive came through the cheers and shouts and most of all the smiles of our fellow Americans. We hope and think that those people shared our belief that this is the beginning of a new era, the beginning of an era when man understands the universe around him, and the beginning of the era when man understands himself.[16]

Eagle's ladder and plaque: "We came in peace for all mankind."
Photo credit: NASA AS11-40-5897.

---

[16] "California Dinner for Apollo 11 Astronauts, part 2," Audio, 11:38, accessed January 15, 2020.
https://archive.org/details/CaliforniaDinnerForApollo11Astronauts/California+Dinner+for+Astronauts_8-13-1969_side+2_.wav.

Task Accomplished. Mission Control erupts with the safe return of Apollo 11.
David Reed is left of center, in the light jacket facing the camera.
Photo credit: NASA S69-40023, Kipp Teague.

Armstrong, Collins, Aldrin, and President Nixon enjoy their reunion aboard the USS Hornet. Photo credit: NASA.

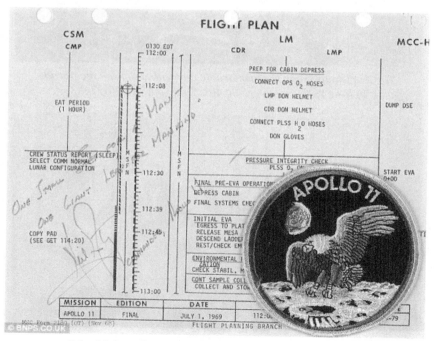

*John McLeaish's flight plan, signed by Neil Armstrong.
Photo credit: UK Daily Mail.*

*The astronauts begin the memorable day with a ticker-tape parade
in New York. Photo credit: Bill Taub, NASA S70-17433.*

*Richard Nixon presents the Presidential Medal of Freedom to the Apollo 11
Astronauts during an emotional evening in Los Angeles.
Photo credit: Richard Nixon Foundation.*

24

GREAT JOB, GENERAL SAM

THE UNITED STATES CONGRESS, SEPTEMBER 16, 1969

"America's Moon astronauts received a tumultuous reception today as they appeared before a joint meeting of Congress," wrote Marjorie Hunter of the *New York Times*. "There were cheers and thunderous applause as the three astronauts walked down the aisle of a packed House chamber to report on their voyage to the Moon."[1] The Speaker of the House of Representatives, John McCormack of Massachusetts, introduced the astronauts, whose speeches were often interrupted by enthusiastic ovations.

> MR. ARMSTRONG: Mr. Speaker, Mr. President, Members of Congress, distinguished guests, we are greatly honored that you have invited us here today. Only now have we completed our journey to land on and explore the Moon and return. It was here in these Halls that our venture really began. Here the Space Act of 1958 was framed, the chartering document of the National Aeronautics and Space Administration. And here, in the years that followed, the key decisions that permitted the successive steps of Mercury and Gemini and Apollo were permitted.
>
> Your policies and the marvels of modern communication have permitted people around the world to share the excitement of our exploration. And, although you have been

---

[1] Marjorie Hunter, "Moon Astronauts Cheered in Congress," *New York Times*, September 17, 1969, https://timesmachine.nytimes.com.

informed of the results of Apollo 11, we are particularly pleased to have this opportunity to complete our work by reporting to you and through you to the American people. My colleagues share the honor of presenting this report. First, it is my pleasure to present Col. Edwin Aldrin.

COLONEL ALDRIN: Distinguished ladies and gentlemen, it is with a great sense of pride as an American and with humility as a human being that I say to you today what no men have been privileged to say before: "We walked on the Moon." But the footprints at Tranquility Base belong to more than the crew of Apollo 11. They were put there by hundreds of thousands of people across this country, people in Government, industry, and universities, the teams and crews that preceded us, all who strived throughout the years with Mercury, Gemini, and Apollo. Those footprints belong to the American people and you, their representatives, who accepted and supported the inevitable challenge of the Moon. And, since we came in peace for all mankind, those footprints belong also to all people of the world. As the Moon shines impartially on all those looking up from our spinning Earth, so do we hope the benefits of space exploration will be spread equally with a harmonizing influence to all mankind.

Our steps in space have been a symbol of this country's way of life as we open our doors and windows to the world to view our successes and failures and as we share with all nations our discovery. What this country does with the lessons of Apollo apply to domestic problems, and what we do in further space exploration programs will determine just how giant a leap we have taken.

COLONEL COLLINS: Many years before there was a space program, my father had a favorite quotation: "He who would bring back the wealth of the Indies must take the wealth of the Indies with him." This we have done. We have taken to the Moon the wealth of this nation, the vision of its political leaders, the intelligence of its scientists, the dedication of its engineers, the careful craftsmanship of its workers, and the enthusiastic support of its people. We have brought back rocks. And I think it is a fair trade. [Laughter erupted from the assembly.]

We could look toward the Moon, toward Mars, toward
our future in space—toward the new Indies—or we could
look back toward the Earth, our home, with its problems
spawned over more than a millennium of human occupancy.
We looked both ways. We saw both, and I think that is what
our nation must do. Man has always gone where he has been
able to go. It is that simple. He will continue pushing back
his frontier, no matter how far it may carry him from his
homeland.

Someday in the not-too-distant future, when I listen to an
earthling step out onto the surface of Mars or some other
planet, just as I listened to Neil step out onto the surface of
the Moon, I hope I hear him say: "I come from the United
States of America."

MR. ARMSTRONG: We landed on the Sea of Tranquility in the
cool of the early lunar morning when the long shadows
would aid our perception. The Sun was only 10 degrees
above the horizon. While the Earth turned through nearly a
full day during our stay, the Sun at Tranquility Base rose
barely 11 degrees—a small fraction of the month-long lunar
day. There was a peculiar sensation of the duality of time—
the swift rush of events that characterizes all our lives and
the ponderous parade which marks the aging of the universe.

Both kinds of time were evident: the first, by the routine
events of the flight, whose planning and execution were
detailed to fractions of a second, the latter by rocks around
us, unchanged throughout the history of man, whose three-
billion-year-old secrets made them the treasure we sought.

The plaque on the *Eagle,* which summarized our hopes,
bears this message: "Here men from the planet Earth first set
foot upon the Moon, July 1969 A.D. We came in peace for
all mankind."

I stood in the highlands of this nation near the Continental
Divide, introducing to my sons the wonders of nature and
pleasures of looking for deer and for elk. In their enthusiasm
for the view, they frequently stumbled on the rocky trails, but
when they looked only to their footing, they did not see the
elk. To those of you who have advocated looking high, we
owe our sincere gratitude, for you have granted us the
opportunity to see some of the grandest views of the Creator.

To those of you who have been our honest critics, we also thank, for you have reminded us that we dare not forget to watch the trail. We carried on Apollo 11 two flags of this Union that had flown over the Capitol, one over the House of Representatives, one over the Senate. It is our privilege to return them now in these Halls which exemplify man's highest purpose—to serve one's fellow man.

We thank you, on behalf of all the men of Apollo, for giving us the privilege of joining you in serving—for all mankind.[2]

The wives of the three astronauts watched and cheered from the President's balcony. After the 30-minute ceremony, NASA Administrator Dr. Thomas Paine presented a Moon rock to the Smithsonian Institution.[3]

WORLD TOUR

On September 29, the crew of Apollo 11 and their wives embarked on a 38-day, 29-stop tour covering 22 nations. *New York Times* reporter Nan Robertson wrote, "The three Apollo 11 astronauts provided living proof today that a world tour can take more out of a man than a trip half a million miles to the Moon and back."[4]

Following the expedition, President Nixon greeted the astronauts and their wives with another hero's welcome at the White House, calling their journey "the most successful goodwill trip in the history of the United States." Nixon added, "As the first men ever to land on the Moon you have demonstrated that you are the best possible ambassadors for peace here on Earth."

Armstrong responded they had been received "not just as individuals, but as representatives of the United States, representatives

---

[2] "Transcript of Astronauts' Address to Congress," *New York Times*, September 17, 1969, https://timesmachine.nytimes.com.

[3] Marjorie Hunter, "Moon Astronauts Cheered in Congress," *New York Times*.

[4] Nan Robertson, "Apollo 11 Crew Feted by Nixons On Returning from World Tour," *New York Times*, November 6, 1969, https://timesmachine.nytimes.com.

of a scientific and technological accomplishment—a symbol of a nation firm in its will to share for the benefit of all mankind."[5]

The astronauts and wives spent a quiet and restful evening at the White House. The next day they returned to Houston to brief the Apollo 12 crew before their November 14 launch.

SAM PHILLIPS

The triumphant return of Apollo 11 fulfilled President Kennedy's goal and completed General Sam Phillips' mission at NASA. Soon he would return to the Air Force. His initial concern about being forgotten at the civilian space agency did not materialize; during his years at NASA he received two promotions, departing as a Lieutenant General (three-star).[6]

Phillips and his family drove up to the NASA Headquarters in Washington for an Apollo 11 staff party and were surprised to see a large banner high on the building. It simply read, "GREAT JOB, GEN. SAM!" At the party, he was presented with this note:

GENERAL PHILLIPS –

We, your staff, are most honored to greet you in recognition of your having led our nation, and mankind, to its greatest achievement. We honor you for many reasons; much has already been written, much has already been said. We agree with all of those superlatives which have been used to describe your successful leadership of the Apollo Program: super manager, calmly rational, and all the rest. Yes, even the lines about your being zero on small talk.

But, as mankind presses forward in space, and this landmark achievement takes its place in history, we of the Apollo Program Staff will remember well some things which have not been said to the world. Not said because only those standing before you now are among the privileged few who know Sam Phillips, the man; how he has contributed to the growth of each of us by giving unstintingly of the appreciative

---

[5] Nan Robertson, "Apollo 11 Crew Feted by Nixons On Returning from World Tour," *New York Times*, November 6, 1969.
[6] John Noble Wilford, *We Reach the Moon*, (New York: Bantam Books, 1969), 228.

word, the pat on the back, the calm, gentle, yet firm and unconfused direction. You give your all, and by so doing have brought out the best in all of us. This is what we will remember: You have led a superhuman effort without losing your humanity. You have done a most difficult job with humility, with style, and with grace. We, your staff, can only revel in our association with you and say,

GREAT JOB, GENERAL SAM![7]

NASA Headquarters in Washington displaying the banner for General Phillips. Photo credit: Kathleen Phillips Esposito.

---

[7] Many thanks to General Phillips' daughter, Kathleen Phillips Esposito, for providing the text of the note and photograph of the Headquarters building. Kathleen Phillips Esposito, email message to author, July 17, 2020.

25

AFTER THE MOON LANDING

"I believe the lunar landing changed forever the way all mankind thought about the Moon. I doubt that any following generation will have the same mystical, romantic view of the Moon as the pre-Apollo generations." – Doug Ward [1]

EMIL SCHIESSER

After experiencing the successful landing of Apollo 11 from the back row of Mission Control, General Sam Phillips told Bill Tindall, "On the next mission, I want a pinpoint landing." [2] It was necessary not only to orchestrate an efficient rendezvous but also for the benefit of science. Upcoming missions had to land in tight spaces to explore areas of scientific interest. NASA decided Apollo 12 should touch down within walking distance of Surveyor 3, an unmanned spacecraft that landed on a crater's slope in the Ocean of Storms a few years earlier. In addition to proving they could hit a small target, the astronauts would also return

---

[1] Doug Ward, email message to author, April 5, 2019.
[2] Murray and Cox, *Apollo*, 383.

with a piece of Surveyor 3 so engineers could study the long-term effects of the lunar environment on equipment.

Bill Tindall met with his group to discuss ways to make it happen. In a classic Tindallgram named "How to land next to a Surveyor: a short novel for do-it-yourselfers," Tindall convened "a three-day Mission Techniques free-for-all starting July 30 to see what we could jury-rig together to improve our chances of landing next to the Surveyor."[3] From the meeting came a solution from a brilliant mathematician named Emil Schiesser.[4]

Schiesser devised a way to determine from tracking data where the LM autopilot was trying to land based on its position where the powered descent began. Apollo 11 landed a few miles downrange because its descent started downrange. Schiesser's calculations would use tracking tools such as Doppler radar to determine the precise location where Apollo 12's Powered Descent Initiation actually occurred compared to its planned location. Once the difference was calculated, the onboard computer would be told that the landing site had changed by the equivalent amount. This would fool the computer, causing it to offset the error.[5]

Schiesser brought up the idea earlier in the year during a Mission Techniques meeting on April 9. It was rejected because of possible complications in the rendezvous if there was an abort during the descent.[6]

David Reed was FDO for the Apollo 12 descent. He met with Commander Pete Conrad to describe the strategy. Reed asked Conrad where he wanted to land, and Conrad indicated the far side of the Surveyor Crater. "Then one day Conrad told Reed he didn't want to walk that far," said Murray and Cox, who describe the conversation:

---

[3] Bill Tindall, "How to land next to a Surveyor: a short novel for do-it-yourselfers."
[4] After Neil Armstrong retired from NASA, a historian asked who at the Manned Space Center impressed him the most. Armstrong answered Emil Schiesser. "I'd vote for Emil every time." Murray and Cox, *Apollo*, 386.
[5] For a detailed description, *see* Emil Schiesser's email to David Reed, which Mr. Reed has graciously provided for Appendix C, pages 339–341.
[6] Bill Tindall, "How the MSFN and sextant data are used to target DOI and Descent," Tindallgram, April 16, 1969, http://www.collectspace.com/resources/tindallgrams/tindallgrams02.pdf, David Reed email message to author, March 16, 2021.

"Put me on the near edge of the crater," Conrad said. So, Reed changed the parameters in the software. Watching Reed calculate these minute changes, apparently in all seriousness—the crater itself was only 600 feet across—Conrad finally told Reed, "You can't hit it anyhow! Target me for the Surveyor." "You got it, babe," Reed said, and he proceeded to enter coordinates that would land the LM precisely on top of the Surveyor if Schiesser's idea worked.[7]

Other improvements emerged for future Apollo mission planning. Experience indicated tracking was more accurate if the LM made two consecutive revolutions separated from the CM. Tindall wrote in an August 29, 1969 memo, "This leads to the first proposal, which David Reed has been pushing for a long time, in spite of our ignorance. Namely, undock one rev earlier so that we can get two complete tracking passes on the LM alone."[8] This adjustment would also allow the undocked LM to perform two separate star checks using their Alignment Optical Telescope, decreasing the angst Steve Bales endured before the Apollo 11 descent.

And then there was the issue of venting. The Apollo 11 crew failed to remove all the air out of the docking tunnel before the LM separated from the CM, which contributed to the powered descent beginning a few miles downrange. Gene Kranz compares pressure remaining in the tunnel to popping a cork on a champagne bottle. When the two spacecraft undocked, force added a slight amount of thrust to the LM, which became significant over the eighty minutes until PDI.[9]

It alerted people to the unwanted consequences of seemingly insignificant venting. Bill Tindall called for a detailed study of all possible venting sources affecting the spacecraft's propulsion. In his August 28, 1969 Tindallgram titled "Vent, bent, descent, lament!" he wrote, "I just cannot believe that in this day of technological achievement we have to put up with spacecraft which make maneuvers we don't want. Until now, this characteristic has been primarily an

---

[7] Murray and Cox, *Apollo*, 385.
[8] Bill Tindall, "A lengthy status report on lunar point landing including some remarks about CSM DOI," August 29, 1969,
http://www.collectspace.com/resources/tindallgrams/tindallgrams02.pdf.
[9] Kranz, *Failure Is Not an Option*, 282.

annoyance, but when it comes to attempting pinpoint landing on the lunar surface, it can really tear us up!"[10]

In another memo, he noted the effects of a water boiler on the Lunar Module.[11] Initial estimates predicted the water evaporating into space would cause the orbiting LM to have a 4,000-foot error in the point of Powered Descent Initiation unless it was corrected.[12]

On November 19, 1969, Pete Conrad and LMP Alan Bean began their descent to the lunar surface. One minute and fifteen seconds after Powered Descent Initiation, Houston determined PDI occurred 4,200 feet short of the planned point. CAPCOM Jerry Carr gave the crew a targeting adjustment of "plus 4,200," which Al Bean entered into the onboard computer. Sure enough, the error was soon corrected, and the Lunar Module was on target for the rest of the descent.

After pitchover, Pete Conrad took his first glance at the landing site and shouted, "Hey—there it is! There it is! Son of a gun! Right down the middle of the road! It's targeted right for the center of the crater!"

As Bean called out the altitude, he said, "The boys on the ground do okay." Conrad flew over Surveyor Crater and landed nearby. The pinpoint landing was a complete success. In addition to collecting valuable lunar samples, the crew removed part of the Surveyor and brought it back for inspection.

However, the crew made a critical mistake early in the first of two moonwalks. They inadvertently pointed the Westinghouse color television camera toward the Sun, burning out a sensitive tube and rendering the camera useless for the rest of the mission. It ended live television coverage from the Moon's surface on Apollo 12, and sadly, of all future moonwalks for United States viewers. The American networks showed the Apollo 11 EVA from beginning to end and planned to do the same for Apollo 12. It was an expensive proposition, but they felt compelled to do so.

---

[10] Bill Tindall, "Vent, bent, descent, lament," August 28, 1969, https://finleyquality.net/wp-content/uploads/2019/07/HSI-39249-Tindall-1969-08-28-Vent-bent-descent-lament.pdf.

[11] Bill Tindall, "Apollo 12 Descent—Final Comments," November 4, 1969, http://www.collectspace.com/resources/tindallgrams/tindallgrams02.pdf.

[12] Bill Tindall, "Apollo 12 Descent—Final Comments." When you consider the LM is orbiting at over 5,200 feet per second, a 4,000-foot error equates to a velocity change of less than one foot per second during the two hours from undocking to PDI.

Doug Ward said, "The networks did the live broadcasts out of a competitive sense and a sense of responsibility, but economically it made no sense to them. And when they saw on Apollo 12 that they could get away without it, from then on, they covered missions in a way that was most economical for them. The public did not get to see in real time all of the amazing lunar surface activity." Ward continued,

> The television that we got from Apollo 15, 16, and 17 were some of the most spectacular space television ever. It was just really great stuff. They had an improved RCA camera, and it did not require much of the astronauts' time or participation to make it work. Ed Fendell was able to control it from the Control Center. And it was the perfect setup for watching everything the astronauts did. It worked really well and was an excellent picture. But the American networks covered almost none of it live. They would edit pieces of it and use it on the evening news, but I think the camera failure on Apollo 12 was what set that up.[13]

PARIS AIR SHOW, 1989

After his years with NASA, Jerry Bostick worked at Grumman Aircraft Engineering Corporation when the Apollo 11 crew was honored at the 1989 Paris Air Show. NASA Public Affairs asked Bostick to shelter the three heroes from the crowds for a day. Bostick took them to the Grumman chalet where they enjoyed many conversations. While standing with Neil Armstrong outside, Bostick asked him what he would have done if Bob Carlton had called a no-fuel abort in the final moments of the descent. Armstrong looked at Bostick, grinned, and replied, "I would have probably hesitated a few seconds and then said, 'Say again, Houston?' And by that time, I would have been on the surface!"[14]

Bostick has no doubt he would have found a way to land. "He was an extraordinary pilot!"

---

[13] Doug Ward, interview by author, March 16, 2019.
[14] Jerry Bostick, interview by author, August 25, 2020.

THE MOON ABOVE, THE EARTH BELOW

In July 1969, while the world focused on the Moon and the three brave explorers aboard Apollo 11, CBS News had the foresight to take a snapshot of the country from which they came. Over 30 camera crews fanned out across America to capture life's ordinary moments on the day man landed on the Moon. The film was archived for 20 years and aired for the first time on July 13, 1989, at 9:00 p.m. Eastern Daylight Time. CBS News called their production "The Moon Above, The Earth Below." Dan Rather stood in front of the Naval Observatory in Washington, introducing the remarkable program:

> In the sixties, it was apparent that the rockets carrying nuclear warheads had become so powerful that either we or the Russians would get to the Moon. President John Kennedy ordered the United States to send a man to the Moon and bring him back safely before the end of that decade. We got there at considerable risk and at great cost. We went to the bright side of the Moon in the dark years of the Cold War. And yet, to my mind at least, nothing in this century has excited us more, united us more than the Moon landing. It was not only an American triumph but the greatest adventure of the 20th century. The story of Apollo 11's voyage to the Moon and its three-man crew is one of great risk, risks far greater than were acknowledged then. If they got there, they'd be heroes. If they failed, they'd be martyrs.[15]

The program interlaced mission audio/video with everyday scenes around the country. On July 20, 1969, Charles Kuralt was assigned to the island of Kauai, the western edge of the United States. He and the camera crew flew there by helicopter. Kuralt stood by the seashore. Behind him were massive cliffs and the setting Sun nearing the horizon. During the latter portion of the moonwalk, filled with awe, he reported, "This is the last light falling on the United States. Those are the Na Pali

---

[15] "The Moon Above, The Earth Below," CBS News, written and produced by Perry Wolff, hosted by Dan Rather and Charles Kuralt, aired July 13, 1989, video, 1:34:39. This highly recommended program can be viewed online in eight segments on Mark Rogers' YouTube page, uploaded July 22, 2011, https://www.youtube.com/watch?v=L995KL1yrx0.

Cliffs on the island of Kauai, westernmost of the Hawaiian Islands. America ends here. Neil Armstrong and Edwin Aldrin right now are standing on the Moon. The rest of us will always remember where we were at this moment."

Twenty years later, Kuralt reflected on scenes around the world showing people looking with wonder at their televisions. "I thought never in human history have so many people shared the same moment, shared the same tongue, no matter what the language of astonishment might be."[16]

Neil Armstrong with the experimental X-15 rocket plane.
Photo credit: NASA DFRC-E60-6286, Kipp Teague.

---

[16] CBS News, "The Moon Above, The Earth Below."

Armstrong in the suit room before Apollo 11 Launch.
Photo credit: NASA KSC-69PC-377, Kipp Teague.

Bill Tindall led the all-important Mission Techniques meetings,
synthesizing information from subject matter experts
and finding the best way to make it work. Photo credit: NASA.

THE MISSION CONTINUES

EX LUNA, SCIENTIA

After the successful missions of Apollo 11 and Apollo 12, the lunar missions took on more of a scientific role. The Apollo 13 crew chose the Latin phrase, *Ex Luna, Scientia*, "From the Moon, Knowledge," for their mission patch.

Apollo 13 launched at 13:13 military time, or 1:13 p.m., on Saturday, April 11, 1970. By Monday night, the mission was going so smoothly that CAPCOM Joe Kerwin told Commander Jim Lovell, "The spacecraft is in real good shape as far as we're concerned, Jim. We're bored to tears down here." It looked like the smoothest Apollo flight to date as they prepared for an evening television broadcast.

Steve Bales was visiting Guidance Officer Will Fenner on the front row of Mission Control to see how the mission was progressing.[1] On the row behind Fenner, Sy Liebergot manned the EECOM (Electrical, Environmental, and Consumables) console monitoring the electrical and life-support systems for the Command/Service Module.

The Apollo Service Module carried super-cooled hydrogen and oxygen for its fuel cells. This allowed higher amounts to be carried on the lunar voyage, but it complicated the process of measuring their quantities. Each element stratified into layers in their cryogenic tanks, much like a milkshake left standing. To get accurate quantity readings, they had to be stirred daily by a fan in each tank. Liebergot noticed an irregular reading after a cryo tank stir a few hours earlier and wanted to ensure it was normal before handing off to the next team of flight controllers. He requested another cryo stir after the broadcast.

---

[1] Steve Bales, email message to author, October 13, 2020. Bales did not have a flight control assignment because he began preparations for Skylab after Apollo 12.

A few seconds later, Will Fenner and Steve Bales noticed the spacecraft's onboard computer rebooted. Initially, they suspected a hardware or software issue.[2]

Fenner calmly reported to Flight Director Gene Kranz, "We've had a hardware restart. I don't know what it was."

Then Kranz heard Jim Lovell's famous words, "Houston, we've had a problem."

Within seconds, noise filled the previously-quiet MOCR as controllers anxiously talked to Kranz and their back rooms. Data on their displays indicated multiple failures, but nobody in Mission Control could believe something so extreme was possible. The quiet mission took a sudden turn of epic proportions.

For the rest of the story, please see
Book Two: Apollo 16's Journey to the Lunar Highlands.

*Flight Director Gene Kranz (back row, below screen) views Fred Haise during the Apollo 13 telecast just before the explosion.
Photo credit: NASA S70-35139, Kipp Teague.*

---

[2] Later they learned the explosion jolted the computer's electrical supply.
Steve Bales, email message to author, April 15, 2020.

Appendix A:
Apollo Glossary

AOT (Alignment Optical Telescope). A sighting scope operated manually by astronauts in the Lunar Module.

Apogee. The high point of an orbit.

Apollo. Project name for the manned lunar landing program, also the spacecraft involved with manned lunar landing program.

ARIA (Apollo Range Instrumentation Aircraft). Eight C-135A jet aircraft were modified to provide mobile telemetry and communications with spacecraft which are out of range of tracking stations. Renamed Advanced Range Instrumentation Aircraft after the end of Apollo.

Boilerplate. A term derived from steel manufacturing, referring to a test article or mock-up. For example, a capsule with similar size, shape and mass but without functioning systems.

Bus. A distribution point supplying power to most of the spacecraft systems.

Cabin. The inside of a spacecraft where the crew operates.

CAPCOM. Capsule Communicator in Mission Control, an interface between Mission Control and the astronauts, call sign is "Houston."

CDR. See Commander.

Circularization maneuver (Circ). The spacecraft adjusts its elliptical orbit to a circular one.

CMP. See Command Module Pilot.

Command Module (CM). A cone-shaped spacecraft that serves as a mother ship orbiting the Moon while the Lunar Module lands. The CM is 11.5 feet high with a base 13 feet in diameter. It is protected by a heat shield to withstand the high temperatures of entry.

Command Module Pilot (CMP). The CMP flies the Command Module and remains in the CM orbiting the Moon while the Commander and Lunar Module Pilot explore the lunar surface. The CMP performs most major maneuvers in the Command Module not involving the launch.

Commander (CDR). The leader of the flight crew. In Apollo, the CDR flies the Lunar Module and is the first out of the hatch for lunar surface EVAs.

Command/Service Module (CSM). A combination of the cone-shaped Command Module and cylinder-shaped Service Module. The Service Module contains a large engine and supplies power and water to the CM. See Command Module and Service Module.

Control Center. See Mission Control, Mission Control Center, Mission Operations Control Room.

CSM. See Command/Service Module.

Delta-H. A difference in height. During the powered descent, this refers to the difference between altitude readings from the Lunar Module landing radar and the onboard guidance computer.

DOI (Descent Orbit Insertion). A maneuver to adjust the spacecraft's orbit around the Moon so its low point is the desired starting point for the Lunar Module's powered descent to the surface.

DSE (Data Storage Equipment). The Command Module "Black Box" which records voice conversations and telemetry when the spacecraft is on the far side of the Moon and out of radio contact with Earth. The information can be downlinked when communication is reestablished.

DSEA (Data Storage Electronics Assembly). The Lunar Module "Black Box" which records voice conversations and telemetry when the spacecraft is on the far side of the Moon and out of radio contact with Earth. The information can be downlinked when communication is reestablished. Similar to the Command Module's DSE.

DSKY (Display and Keyboard Assembly), rhymes with "risky." The Apollo Guidance Computer user interface with a display and keyboard.

EECOM (Electrical, Environmental, and Communications). The flight controller who monitors the Command/Service Module's electrical and environmental control systems.

Entry. Also called reentry, when a spacecraft negotiates the high heat and G forces of Earth's atmosphere before splashdown.

EVA (Extra Vehicular Activity). Activity performed by an astronaut outside a spacecraft, including spacewalks and moonwalks.

Flight Director (Flight). The boss of a flight controller team in Mission Control who can take any action necessary for mission success.

Flight Dynamics Officer (FDO), pronounced "FIDO." The flight controller responsible for the flight path or trajectory of the spacecraft.

Flight plan. A minute-by-minute timeline of activities scheduled for the flight crew.

Gimbal. A mount or support allowing an object to pivot around an axis. Rocket engines are often mounted by two gimbals so they can swivel on two axes. The Apollo guidance platform is mounted by three gimbals allowing the platform to remain stable in relation to a horizon.

Gimbal lock. Occurs when two of the gimbal rings are parallel to each other, causing the mounted object to lose its freedom of motion.

Guidance Officer (GUIDO, or Guidance). The flight controller monitoring the onboard guidance computer and the position of the spacecraft as calculated by the tracking stations.

Hasselblad. A high-quality medium format camera using 70mm film (as opposed to standard 35mm film) allowing for high-resolution photographs.

High-Gain, or steerable antenna. A radio antenna which focuses its radiation pattern in a narrow beam for better signal strength, as compared to an omnidirectional antenna which spreads its signal in all directions. Antenna gain is inversely proportional to beam width.

Houston. The location of the Manned Spacecraft Center and Mission Control, also the call sign for the Capsule Communicator (CAPCOM).

Instrument Unit (IU). Located above the Saturn third stage, the IU contains a computer and guidance platform for the launch vehicle.

Jettison. Disposal of an unnecessary object.

KSC (Kennedy Space Center). The NASA field center on Merritt Island, Florida, near Orlando, where rockets are prepared and launched.

LAM (Landing Area Map). A 1:100,000 chart used to identify specific locations in the landing area. A grid marked with letters on the left side and numbers along the bottom provides a way to designate any feature on the map. LAM-2 refers to the Landing Area Map for Landing Site 2.

Launch Vehicle. The Saturn V launch vehicle consists of the first (lower) three stages of the rocket plus the Instrument Unit. The launch vehicle is distinguished from the "spacecraft," which usually consists of the Command/Service Module (CSM) and Lunar Module (LM). See Spacecraft.

Limb. The edge of the Moon as viewed from Earth. Radio communications are blocked once the spacecraft goes behind the Moon and loses line of sight with Earth.

LMP. See Lunar Module Pilot.

LOI (Lunar Orbit Insertion). A maneuver slowing a spacecraft to the optimal velocity for lunar orbit.

LOS (Loss of Signal). A break in line-of-sight communications with Earth when a spacecraft goes behind the Moon.

LRL (Lunar Receiving Laboratory). A facility at the Manned Spacecraft Center in Houston which quarantined astronauts and material brought back from the Moon on Apollo 11, 12, and 14.

Lunar Module (LM). The landing module which descends to the lunar surface and ascends to rendezvous with the Command Module.

Lunar Module Pilot (LMP). Monitors the guidance computer and calls out critical readings while the Commander flies the Lunar Module.

Marshall Space Flight Center, MSFC. The NASA field center located near Huntsville, Alabama. NASA's propulsion center which provides Saturn launch vehicles under the leadership of Wernher von Braun.

Massachusetts Institute of Technology, MIT. A private research university in Cambridge, Massachusetts, which excels in the development of technology and science.

Mission Control, or Mission Control Center, MCC. Located in Building 30 on the campus of the Manned Spacecraft Center (now named Johnson Space Center) where flights are monitored and controlled.

Mission Techniques. A group led by Bill Tindall coordinating data and developing specific details of flying a lunar mission.

MIT. See Massachusetts Institute of Technology.

MOCR, Mission Operations Control Room, rhymes with "joker." Commonly known as Mission Control or the Control Center. There were two identical MOCRs in Building 30. MOCR-2 on the third floor was the nerve center for manned Apollo flights and is now designated a National Historic Landmark. MOCR-1 on the second floor served a variety of purposes, such as monitoring unmanned Apollo test flights, training simulations, and operating Skylab missions. Both MOCRs were used for Shuttle flights.

MSC (Manned Spacecraft Center). The NASA field center in Houston, Texas, responsible for the development and operation of space flight. It was renamed Johnson Space Center in 1973.

MSFC. See Marshall Space Flight Center.

MSFN (Manned Space Flight Network), pronounced "misfin." A network of tracking stations supporting Mercury, Gemini, Apollo, and Skylab missions. Large dish antennas were incorporated for tracking Apollo missions to the Moon.

NASA (National Aeronautics and Space Administration). Succeeded the National Advisory Committee for Aeronautics (NACA) in 1958 and pursues the peaceful exploration of space.

Network. The flight controller who coordinates the ground tracking stations.

Nominal. Normal, according to plan, within expected or acceptable limits.

Omni, or Omnidirectional Antenna. A radio antenna which radiates in all directions, as compared to a high-gain steerable antenna which has a higher signal strength in a narrow direction. Antenna gain is inversely proportional to beam width.

OPS (Oxygen Purge System). Provides 30 minutes of emergency oxygen in case of a failure in the Portable Life Support System (PLSS) backpacks during a moonwalk.

Orbit. The movement of an object around a celestial body.

P00, (Program Zero-Zero), pronounced "Pooh." Places the Apollo Guidance Computer in an idle state, where it can be controlled by Houston.

PAD (Pre-Advisory Data). A series of numbers describing an upcoming burn, indicating details like the time of ignition, duration, and expected change of velocity.

PAO (Public Affairs Officer). Provides public relations services. During missions, PAOs perform mission commentary so the media can understand details of the flight and report accurately.

Passive Thermal Control (PTC, or barbeque mode). In order to balance the heat from the Sun with the cold of space, the spacecraft flies to and from the Moon with its long axis perpendicular to the Sun, rolling three times per hour.

PDI. See Powered Descent Initiation.

Pericynthion. The point nearest the Moon of an object in lunar orbit, where the object originates from somewhere else, like Earth.

Perigee. The low point of an orbit.

Perilune. The low point of an orbit around the Moon, similar to pericynthion.

Pitch. The rotation of an airplane causing the nose to move up or down.

PLSS (Portable Life Support System), rhymes with "bliss." Backpacks worn my moonwalkers which supply the astronauts with oxygen, water, and a comfortable environment.

Powered Descent. The landing phase of a lunar mission. The Lunar Module descent engine thrusts against the direction of travel and acts like a brake. The throttleable engine adjusts to the decreasing weight and velocity of the Lunar Module as it approaches the Moon.

Powered Descent Initiation, PDI. The beginning of the powered descent, where the Lunar Module descent engine is ignited against the direction of travel in a braking maneuver.

RCS (Reaction Control System, thrusters). Provides small amounts of thrust and velocity change in any direction for fine-tuning or adjusting the orientation in space.

Reentry. Another name for entry, when the spacecraft negotiates the high heat and G forces of Earth's atmosphere before splashdown.

Retrofire Officer (RETRO). The flight controller responsible for getting the spacecraft home, whether during an abort or at the end of a nominal (normal) mission.

Retrograde burn. Against the direction of travel.

Revolution. A rotation of an object around a fixed point, often used synonymously with "orbit."

Roll. The rotation around the long axis of an airplane where the wings move up and down.

Service Module (SM). A cylinder-shaped module mated to the conical-shaped Command Module. The SM is 15 feet long and 13 feet in diameter and contains a large Service Propulsion System (SPS) engine to propel the spacecraft back to Earth from the Moon. It also contains fuel cells to power the spacecraft and the hydrogen and oxygen needed as reactants.

Solar Wind. A stream of particles released from the Sun's corona.

Spacecraft. Most often used in reference to the Command/Service Module (CSM) and Lunar Module (LM). Early in the mission, it also includes the Spacecraft-Lunar Module Adapter (SLA) and the Escape Tower. "Spacecraft" is distinguished from the Saturn launch vehicle. See Launch Vehicle.

SPS (Service Propulsion System). The powerful Service Module engine performing maneuvers near the Moon, including the Trans-Earth Injection to return the crews home.

SRC (Sample Return Container) or rock box. Rectangular aluminum box, triple-sealed, in which the Moon rocks are transported to the Lunar Receiving Laboratory in Houston.

SSR (Staff Support Room) or "back room." Rooms near the Mission Control Center filled with experts to support flight controllers.

State vector. A group of numbers telling the spacecraft where it is, in what direction it is heading, and how fast it is moving.

Station-keeping. Two spacecraft orbiting in close formation, oriented where they can observe each other.

Steerable antenna. See High-Gain antenna.

TEI (Trans-Earth Injection). A maneuver using the Service Module engine (SPS) to send the spacecraft out of lunar orbit and on a trajectory to intersect Earth.

Telemetry. The process of taking measurements via sensors, converting the data to electrical signals which are transmitted by radio to tracking stations on Earth, and then decoded back into the original data.

TLI (Translunar Injection). A maneuver using the third stage engine to send the spacecraft out of Earth orbit to the Moon.

Tracking stations. Strategically located around the world, these stations provide communications with the spacecraft using antennas, sensitive receivers, and lines of communication with other facilities in a network. See MSFN.

Trench. The first row of flight controllers in Mission Control where the trajectory specialists operate.

Ullage burn. A burn performed by thrusters just before main engine ignition to force propellants into the combustion chamber. This is similar to momentum pushing passengers forward as a car slows.

VOX (Voice Activated Transmission). A mode of communication where one's voice activates transmission, as opposed to "push-to-talk" mode.

Yaw. The rotation of an airplane where the nose moves left and right.

Honeysuckle Creek Tracking Station, Australia. Photo credit: John Saxon.

Neil Armstrong (left) and Colin Mackellar in Sydney, Australia, 2011.
Photo credit: Colin Mackellar.

Appendix B:
The Hunt for Apollo 11 Tapes

John Sarkissian is a former Operations Scientist at the Parkes Radio Observatory in Australia. He wrote an article for the Australian Physics magazine in the summer of 2012 describing the search for missing high-quality tapes of the Apollo 11 moonwalk (link below). Honeysuckle Creek historian Colin Mackellar was also on the search team and has graciously provided his insights. This appendix is a summary of both of their contributions.

The images of Neil Armstrong's first steps upon the Moon are unforgettable. Like many aspects of Apollo, the process by which we witnessed the event on live television was exceptional.

The camera was made by Westinghouse under the leadership of Stan Lebar. It transmitted slow-scan television (SSTV) signals to Earth, requiring much less power and bandwidth than normal television. Three tracking stations received the signals during the moonwalk. Goldstone in California and Parkes in Australia had 64-meter dishes, while Honeysuckle Creek in Australia had a 26-meter dish. The black-and-white slow-scan video was sent from the Moon at 10 frames per second with 320 lines per frame. Local RCA scan converters adjusted them to US commercial television standards of 30 frames per second with 525 lines per frame.

The conversion process was simple but effective. Each frame received from the Moon appeared on a high-resolution CRT (Cathode Ray Tube) monitor inside the scan converter. A Vidicon camera was aimed at the screen and imaged each frame at commercial television standards. The resulting signals were sent to Houston where a technician would select the best-quality images for worldwide television. The conversion process added signal noise, degraded the image, and caused ghosting when bright images moved across the screen. The process of relaying the signals to Houston via satellite scrubbed more quality. Thus technicians at the tracking stations saw much clearer images than the rest of the world.

The unconverted video was one of 15 tracks recorded on one-inch magnetic telemetry tapes at the three tracking stations. The other 14 tracks included voice, biomedical readings, EVA backpack systems information, and other data. The huge telemetry tapes were sent to NASA after the mission. The scan-converted images in commercial television standards were properly recorded on black-and-white film and archived for history.

As Armstrong descended the ladder, the television from Goldstone was dark with a high degree of contrast. Despite their smaller dish, Honeysuckle Creek had better images and thus provided the historic first steps to the watching world. Later, the Parkes main beam received signals once the Moon rose 30 degrees above the horizon, resulting in the best pictures for the remainder of the moonwalk. Technicians switched to the Parkes feed a few minutes into the EVA.

John Sarkissian began working at Parkes in 1996 and wanted to see the original SSTV images received from the Moon before they were converted and degraded. His interest was piqued in 2001 when he saw a polaroid photo taken of the slow-scan monitor during the moonwalk; it was of much higher quality than anything seen on television. The hunt for the original recordings intensified, and a search team emerged. Joining Sarkissian was Stan Lebar (Westinghouse engineer who was the program manager for the Apollo Lunar Surface Camera), Dick Nafzger (Goddard Space Flight Center engineer responsible for supporting Apollo TV), Bill Wood (lead engineer at Goldstone and was on the console during the moonwalk), and Colin Mackellar (historian for the Honeysuckle Creek tracking station). They left no stone unturned, looking everywhere from the National Archives in Washington, DC, to garages of technicians who worked at the tracking stations.

Meanwhile, another search began at Honeysuckle Creek for a machine to play such tapes if they were located. This was prompted by Operations Supervisor John Saxon and Colin Mackellar learning a former technician may have had a copy of one of the reels but had no machine to play it. Saxon sent over 450 emails to his NASA friends. Dick Nafzger discovered NASA-Goddard had a seven-foot high legacy machine which could play the archaic tape. It worked, but unfortunately the material was from a simulation before Apollo 8. At least they knew the telemetry tapes could be played if found.

Mackellar produced side-by-side comparisons of the images seen at Honeysuckle in contrast to what the world saw on television. The difference was so dramatic that the team's presentations reached the worldwide media and caught the attention of space agency personnel including NASA Administrator Mike Griffin. NASA took great interest and formalized the search for the tapes, making their resources available. Eventually, documents were discovered showing NASA had retrieved all of the 250,000 Apollo-era data tapes from the National Archives and erased them for future use, including those with the Apollo 11 slow-scan video. They were considered backup data tapes of completed missions since few people outside of tracking station personnel were aware of the SSTV process.

The tapes were needed because they were manufactured with whale oil, banned in the mid-1970s. After the ban, manufacturers had to use synthetic oils which were not as durable. Because of "Sticky Shred Syndrome," the tape layers of the synthetic style eventually stuck together and were torn apart during playback. NASA retrieved the whale-oil tapes due to the rising need to store data from satellite programs. It seems like a blunder now, but the telemetry tapes were never intended to be the main source of archived media. Video of the moonwalk in commercial television standards was the major archival priority and was recorded to last for centuries.

The search team changed directions to look for the best quality scan-converted video of the EVA. They digitized those images and archived them properly. The team used the best parts of several different sources to make one video of the entire EVA. Restoration processes could remove some of the degrading of the scan-conversion, although not reaching the clarity of images seen in the raw footage from the Moon. Professional restoration of the whole two and a half hours would be enormously expensive, so they restored a few scenes to give NASA a taste of what the final version could be. NASA funded the $245,000 project in 2009 in honor of the 40th anniversary of Apollo 11. "The restoration involved digitally repairing damaged sections of the recordings, removing noise from the video, correcting for vignetting, stabilizing and brightening the TV picture and other adjustments," said John Sarkissian.

NASA and Goddard Space Flight Center produced three sets of the complete restored video: one for the National Archives in Washington, the second for Johnson Space Center in Houston, and the third for Australia in recognition of the tremendous involvement of their tracking stations. A week after the sets were completed, Neil Armstrong visited Sydney to pay tribute to the Australians who faithfully worked to ensure the success of Apollo 11. He symbolically handed over the video to Dr. Phil Diamond, director of Australia's national space agency.

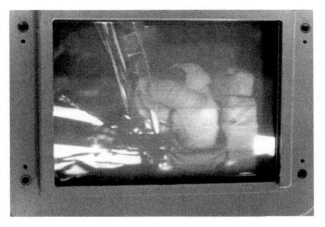

A 35mm photo taken of the scan converter monitor at Honeysuckle Creek (above) compared to the lower-quality converted images received in Houston and sent out to television networks (below).
Credit: Bob Goodman, Colin Mackellar, and honeysucklecreek.net

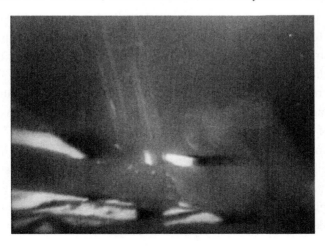

Appendix C:
Emil Schiesser Explains Apollo 12 Pinpoint Landing

This email correspondence from Emil Schiesser to David Reed in April 2011 explains the technical details of how they adjusted the Lunar Module guidance computer's landing point to correspond to the deviation of the Powered Descent Initiation from the planned location. See Chapter 25 for the context. Schiesser refers to his team members including Howard G. DeVezin, W.J. Wylie, and Matt Grogan. Thank you to Mr. Reed for providing this email.

The subjects mentioned in this e-mail include the Apollo 11 landing miss, determination of LM position for Apollo 11 ascent, and Apollo 12 pre-descent navigation and targeting procedures. The following are some comments on two of these: Apollo 11 landing miss and Apollo 12 pre-descent nav.

During Apollo 11 the two-rev forward prediction errors were relatively small, a few thousand feet, through rev 12. By rev 13 the along track error was 7800 ft and at rev 14 PDI over 20,000 feet. The cross-track error was over 4000 feet at PDI. The 4600 ft cross-track landing site miss to the South of the intended location was predominantly due to orbit plane motion prediction error due to the limited lunar gravitation field model. The landing site overshoot, 22,300 feet, included over 1000 ft of crew hazard avoidance. Lacking that, the overshoot would have been around 21,000 feet. I'm pretty sure about the 22,300 ft but not as sure about the hazard avoidance number. Most of the 21,000 ft error occurred on the backside of the Moon prior to landing and was due to unmodeled spacecraft self-induced perturbations which Tindall worked to minimize for Apollo 12. The Apollo 11 LM was barely in the "large-area" landing ellipse (some charts show it barely out). The Apollo 12 "pin-point" landing requirement was to be within a 3000 ft radius of the intended site.

For Apollo 12 the pre-PDI timeline for state vector and landing site target determination and their uplink was changed (from Apollo 11) and means for determining LM along-track position error near PDI, no matter the error source, was embraced along with a way to compensate for it (Noun 69). The Apollo 12 cross-track position error was reduced through the application of two revs of expected orbit plane motion error to landing site target location. The landing site target was computed from downlinked CSM sextant tracking landing site tracking data (line-of-sight angles to a 600 ft crater with known offsets to the desired landing point).

The Apollo 12 along-track error was determined three ways: Use of range data from a MSFN tracker shortly after acquisition of signal (AOS), the use of two-way Doppler data from that site, and use of altitude rate from the Powered Flight Processor (PFP) near PDI. The PFP was a Kalman filter that concurrently used two-way Doppler data from one site and three-way Doppler data from three other sites starting about 8 minutes prior to PDI and through descent and landing.

Range measurements were compared with those generated from a post AOS downlinked LM vector. The differences were adjusted for lunar ephemeris earth-moon distance error, for the deviation of the LM direction of motion relative to the tracker-vehicle line of sight, and for the expected change in the along-track error from there down. The Doppler data, in the form of range-rates, were compared with values computed from the LM downlinked vector (predicted forward). The along track error is geometrically related to the range-rate residual assuming the orbit shape is reasonably known. Altitude-rate from the PFP was compared with that computed from the LM vector. The altitude-rate difference is also geometrically related to the along-track error, given the way the PFP works and assuming the orbit shape is reasonably known. It was necessary to adjust the results obtained from the range-rate and PFP methods using prior rev to rev experience to account for various errors and factors, similar to that mentioned for range. Much of this was done by hand. A calculator was brought in for later flights. The team included DeVezin, Wylie, Grogan, and Flanagan. I was also there. DeVezin and Wylie were in charge of the PFP and sat at the console. Matt Grogan was responsible for the PFP noun 69 final computations on Apollo 12. All are retired but up and around. A 4200 ft along track error obtained from PFP data was chosen for verbal transfer to the LM about 2 minutes into powered descent on Apollo 12. The range and Doppler based range-rate values were 4735 and 4230 ft. The crew entered the 4200 value by hand via Noun 69 which moved the landing site target forward by that amount. Tindall orchestrated the flight techniques including the means for along-track error compensation. Uplink of state vector deltas, a state vector time tag adjustment and of course a whole vector, during powered flight, was considered too risky.

Pre-descent nav procedures changed some between each flight and sometimes had to be adjusted a bit during a flight, but the three "Noun 69" methods, two of which actually did make use of Doppler based measurements, remained in use through Apollo 17. And, per the above, to say that "The LM in that orbit would have a radial rate that would be predictable as well as observable with MSFN doppler radar" is, at its core, ok.

Best wishes.

Emil

Emil Schiesser. Photo credit: NASA.

Apollo 12 Commander Pete Conrad on the LM porch,
about to descend the ladder. Photo credit: NASA AS12-46-6718.

Apollo 12 demonstrated a pinpoint landing on the rim of Surveyor Crater.
Photo credit: NASA AS12-48-7099.

Appendix D:
The Artwork of
Paul and Chris Calle

Paul Calle, the well-known artist who created the 1967 Gemini "Space Twins" stamps, was chosen to design a stamp commemorating the "First Man on The Moon." A series of pencil sketches eventually became the iconic postage stamp.

Calle began work on the project a year before the Apollo 11 mission. NASA sent Calle photographs of the lunar surface training including Armstrong rehearsing the first step. Calle had to guess certain details, like how far the Lunar Module footpads would penetrate the soil, and he got it exactly right.[1]

A week before the Apollo 11 launch, Postmaster General Winton Blount made details of the stamp public, adding that the master die—the source of all plates and stamps—was traveling to the Moon onboard *Eagle*. The crew canceled the first "Moon Letter" as they made the long journey home.[2]

Paul Calle covered Mercury, Gemini, and Apollo programs. NASA officials invited him to be present with the crew the morning of the launch for breakfast and suiting up. He was the only artist allowed that privilege. His historic sketches from the morning have captured the human spirit in a way that photographs and video do not. During breakfast, Michael Collins, also an artist, struck up a conversation with Calle and looked at his sketches.

Soon after the historic mission, Calle began work on a large four-foot by eight-foot oil painting named "The Great Moment." Years later, Michael Collins remarked, "This painting of my friend Neil Armstrong by my friend Paul Calle combines for me the best of two worlds.

---

[1] Calle, Chris, *Celebrating Apollo 11: The Artwork of Paul Calle*, 22–23.
[2] Calle, Chris, *Celebrating Apollo 11: The Artwork of Paul Calle*, 22–23.

NASA's technological achievements and an artist's exquisite interpretations of it."[3]

Chris Calle's book, *Celebrating Apollo 11: The Artwork of Paul Calle* includes sketches and paintings related to the historic mission and stunning artwork from Projects Mercury, Gemini, Apollo, and Apollo-Soyuz. Chris' website is www.callespaceart.com, and you can contact him directly, chris@calleart.com.

Paul Calle's "Accomplishments in Space" stamps were issued in September 1967. It was the first US Postal Service image to cross two stamps.

Postmaster General Winton Blount (right) joins astronauts Mike Collins, Neil Armstrong, and Buzz Aldrin as they dedicate the postage stamp honoring their historic mission. Photo credit: Associated Press.

---

[3] Calle, Chris, *Celebrating Apollo 11: The Artwork of Paul Calle*, (AeroGraphics, Inc,, 2009), 83.

Appendix E:
Ham Radio in Space

Several men involved in the Apollo program shared a passion for amateur (ham) radio and played pivotal roles in connecting amateur radio with the space program. One of the most popular endeavors in the hobby is contacting the International Space Station. Amateur Radio on the International Space Station (ARISS) is an international outreach program for education. It is easy to see the wonderful benefits for students worldwide, but several pioneers had to overcome significant hurdles to make it a reality. A wave of ham radio operators and others sympathetic to the hobby combined forces, including astronauts, Public Affairs Officers, engineers, high-ranking NASA administrators, a JSC Director, and a legendary news reporter.

NBC news correspondent Roy Neal got the ball rolling before Alan Shepard's first manned spaceflight in 1961. Neal was a highly-respected reporter who covered the Space Administration from its infancy through the early Space Shuttle years. He was also an avid amateur radio operator who enjoyed bouncing radio signals off of NASA's Echo communications balloon satellites,[1] and became one of the best-known hams of his generation. Project Mercury's Public Affairs Officer, John "Shorty" Powers, visited Neal's home not long before Shepard's flight. Neal demonstrated to Powers the ability of ham radio to make contacts all around the world, talking to people in Australia, Europe, South America, and South Africa. Powers was intrigued and suggested to NASA that ham radio operators could serve as a backup communications network in emergencies. NASA decided against it.[2]

By 1967, there was a growing interest to integrate amateur radio with the space program. Two Apollo astronauts were licensed ham radio operators, Owen Garriott and Tony England. Garriott was the father of

---

[1] Echo 1, an ancestor of NASA's communication satellites, was launched in August 1960. It had a diameter of 100 feet and orbited one thousand miles above Earth. Echo 2 launched four years later.
[2] Roy Neal, "In Search of The Shuttle: Fun, Frustration, Fatigue," *73 Magazine*, Issue 282, May 1984, 10, https://archive.org/details/73-magazine-1984-03/page/n10/mode/1up?view=theater.

video game developer Richard Garriott, well-known for his role in creating the Ultima series, Shroud of the Avatar, and Tabula Rasa games. In fact, Steve Bales' sons were more impressed by their dad working with Richard Garriott's father than anything Bales did in Mission Control.[3]

Tony England became a ham at the age of 13, and as a tenth grader listened to the beeps of Sputnik overhead on a ham radio he built.[4] England is still an active radio operator, as is Doug Ward, who was instrumental in establishing the ham radio club at Houston's Manned Spacecraft Center (now Johnson Space Center, JSC).

Owen Garriott wanted to operate ham radio on his 1973 Skylab mission, but his request was denied because there was not enough time before launch to make all the necessary changes. As NASA moved into the Shuttle era, a window of opportunity emerged. In 1982 Roy Neal interviewed General James Abramson, Associate Administrator of the Space Shuttle Program, about a shuttle mission taking place. During a break, when the cameras were being repositioned, Neal asked Abramson, "Has anyone ever mentioned ham radio on a space shuttle to you?" Nobody had, but Abramson was interested in hearing what Neal was thinking. After the interview, Neal showed Abramson his transceiver, and they explored the possibilities. Neal suggested a demonstration of amateur radio from an orbiting shuttle would be a great public relations tool. The General said, "I like it, Roy. Why don't you develop a formal proposal? If it makes sense, I'll approve it."[5] Neal immediately called Doug Ward to talk about the details. They decided it would be important to have sponsorship from the American Radio Relay League (ARRL), the national association for amateur radio. Rosalie White at ARRL provided vital support and connected them with the large ham radio community. The Radio Amateur Satellite Corporation (AMSAT) also came on board, partnering with the ARRL in developing the proposal and aiding in technical discussions with NASA.[6]

---

[3] Steve Bales, interview by author, November 6, 2019.
[4] Craig McEwen, "West Fargo Astronaut Recalls 1985 Challenger Launch," *Grand Forks Herald*, July 29, 2015, https://www.grandforksherald.com/news/3807216-west-fargo-astronaut-recalls-1985-challenger-launch.
[5] Neal, "In Search of The Shuttle," 11.
[6] Neal, "In Search of The Shuttle," 11.

Doug Ward also consulted Bob Allnutt, the Associate Deputy Administrator for NASA, with whom he had a close working relationship. Allnutt was a man of action and a key ally in the process. He liked the ham radio idea and said to Ward, "Write me a memo, tell me how you do it and exactly what it would entail, and I'll see if I can make it happen." Ward gathered input from the engineers and operations personnel, wrote the memo, and sent it to Allnutt. Allnutt put his stamp of approval on it and sent it to General Abramson, who also approved it since the public affairs venture would be well worth the agency's time.[7]

Meanwhile, Owen Garriott was training for his 1983 Space Shuttle mission. Again, he was interested in taking ham radio on his flight, but since the spacecraft was already designed and built, there was no cost-effective way to make a penetration through the pressure vessel for an antenna. Garriott came up with the idea of holding a helical antenna up to the Shuttle window, removing the need for an external antenna. Engineer Lou McFadden, also a ham radio operator, designed the antenna and overcame myriads of technical difficulties involved with its construction.[8] "It was about 24 inches in diameter and looked somewhat like a large aluminum cake pan," explained Garriott. "The transceiver then connected to the antenna."[9] McFadden and other engineers tested the radio and antenna with the orbiter *Columbia* to ensure there would be no interference with the other systems in the spacecraft. Those tests were paid for out of NASA's Engineering and Development budget. "It cost around $25,000 to do the electromagnetic compatibility tests with ham radio and the shuttle systems, and they had to do that with the flight vehicle at the Cape," remembers Doug Ward. "They had to send people to the Cape and get all the equipment there and run the tests to make sure it did not interfere with any of the critical systems, and it checked out. And that was the start of ham radio in the spacecraft."[10]

Even with General Abramson and Bob Allnutt on board, two of the highest leaders at NASA, the battle was far from over. Gerry Griffin,

---

[7] Doug Ward, interview by author, March 16, 2019.
[8] Doug Ward, email message to author, June 24, 2021.
[9] NASA, "Ham Radios in Space," August 6, 2008, https://science.nasa.gov/science-news/science-at-nasa/2000/ast19aug_1/.
[10] Doug Ward, interview by author, March 16, 2019.

who had become the Director of the Johnson Space Center, gave his enthusiastic support of the endeavor and overcame several more hurdles.[11] Objections then surfaced from scientists who complained the radio would take away time from their own experiments. Garriott solved the issue by using the radio only in his personal time.

On November 28, 1983, Space Shuttle *Columbia* launched, carrying Owen Garriott and a 2-meter Motorola handheld transceiver, the first ham radio in space. It was the ninth Shuttle mission, also known as Spacelab 1. "When in orbit over land, I could make a CQ, which is a general call, and see who responded," Garriott said. He also arranged contacts with specified times and frequencies. "Among others, I was able to speak with the Amateur Radio Club in my hometown of Enid, Oklahoma, with my mom, with Senator (Barry) Goldwater, and with King Hussein (of Jordan), who was an avid ham."[12] He also spoke to his sons, including Richard, who were at the JSC Radio Club in Houston. It was a Public Relations hit. NASA and the astronauts got behind the idea of Amateur Radio, which had a much easier path going forward.

Tony England flew aboard *Challenger* on the Spacelab 2 mission in July-August 1985. Motorola improved the radio between the two flights because Garriott had difficulty distinguishing the voices he was talking to from other noise. "The Automatic Gain Control (AGC) in Owen's Motorola handheld used signals on adjacent bands as well as the one tuned in," explains England. "With all of the hams on the ground transmitting on those bands, Owen's AGC nearly silenced the channel he was listening to. Thus, the early difficulty communicating. Motorola modified our handheld so that the AGC listened only to the channel the set was tuned to."[13]

Two of England's crewmates, Commander Gordon Fullerton and John-David Bartoe earned their ham license before the flight. They took it to a new level, inventing the name SAREX (Shuttle Amateur Radio Experiment). "The primary goal of SAREX was to spark students' interest in the science, technology, and communications fields by allowing them to talk to Space Shuttle astronauts using amateur radio," said Frank Bauer, AMSAT Vice President for Human

---

[11] Neal, "In Search of The Shuttle," 12.

[12] NASA, "Ham Radios in Space."

[13] Tony England, email message to author, June 25, 2021.

Spaceflight. "In addition, SAREX increased public awareness of NASA's human spaceflight program by permitting radio amateurs worldwide to talk with the Shuttle astronauts."[14]

The JSC Radio Club scheduled contacts with groups of students all around the world for educational purposes before the mission. "We all worked 12 hours on and 12 hours off around the clock and could only use the Amateur Radio during our 12 hours off," England recalls. "By having three crewmen who could use it, we were able to schedule many more contacts. The criterion for a schedule was that the group's intended questions had to demonstrate some knowledge of the science on our flight."[15]

England successfully lobbied to have the capability for Amateur Radio Slow-Scan TV (SSTV) included on the mission. Until then, the Shuttle only received voice and teletype messages. "A picture of Kathi, my wife, was the first picture we received on board," said England. "The Amateur SSTV made Houston understand the value of transmitting images on board."[16] Bauer reports that during the flight, "thousands of hams received and recorded the automated SSTV pictures sent down from *Challenger*, hundreds of hams around the world talked with Tony, and over 6,000 of our nation's youth participated in the ham radio activities on *Challenger* through voice contacts and picture exchange."[17] Again, the radio was also a public relations triumph. SAREX became a long-running educational program sponsored by NASA, ARRL, and AMSAT.

By the time the International Space Station (ISS) was being designed, Doug Ward was working as a Public Affairs Officer in the Space Station program under ISS Program Manager Randy Brinkley. The astronauts were pushing to include amateur radio on the Space Station. Ward advised Brinkley to add a radio and antenna into the original design, and Brinkley liked the idea. Ward recalls, "On the Space Station we were able to do it the right way and get it in as part of the initial plan and put an external antenna on one of the modules. It's been

[14] Frank Bauer, "Amateur Radio on Human Spaceflight Missions – 30 Years," 4, https://www.ariss.org/uploads/1/9/6/8/19681527/human_spaceflight_ham_radio---30_years_rev_h.pdf.
[15] Tony England, email message to author, June 18, 2021.
[16] Tony England, email message to author, June 18, 2021.
[17] Bauer, "Amateur Radio on Human Spaceflight Missions," 5.

a hit ever since."[18] Speaking with an astronaut onboard the ISS is a holy grail for hams. It is much more common for radio operators to bounce signals off the Space Station's repeater and speak to far-away contacts on frequencies that are limited to line-of-sight communications.

Roy Neal continued to play a key role in SAREX and ARISS (Amateur Radio on the ISS) until his death in 2003. In October 2008, Richard Garriott flew aboard the ISS, becoming the first second-generation American in space, following in his father Owen's footsteps. Richard made good use of ARISS, operating a Kenwood SSTV communicator on his mission and sending over 1,000 images from space. In addition, Garriott made over 500 voice contacts with hams around the world.[19]

Owen Garriott with his handheld Motorola radio in 1983.
Photo credit: NASA.

---

[18] Doug Ward, interview by author, March 16, 2019.
[19] Bauer, "Amateur Radio on Human Spaceflight Missions," 18.

Bibliography

Aaron, John W. Interview by Kevin M. Rusnak for NASA Johnson Space Center Oral History Project, January 18, 2000.

"Aeronautics and Space Report of the President, 1972 Activities." Washington, DC: National Aeronautics and Space Council, 1973. https://history.nasa.gov/presrep1972.pdf.

Aldrin, Colonel Edwin E. "Buzz," Jr. *Return to Earth*. With Wayne Warga. New York: Random House, 1973.

Allen, Joseph P. Interview by Jennifer Ross-Nazzal for NASA Johnson Space Center Oral History Project, January 28, 2003.

Allton, Judith. "Catalogue of Apollo Lunar Surface Geological Sampling Tools and Containers." Houston: Lockheed Engineering and Sciences Company for NASA JSC, 1989. https://www.hq.nasa.gov/alsj/tools/Welcome.html.

Anders, William A. Interview by Paul Rollins for NASA Johnson Space Center Oral History Project, October 8, 1997.

Apollo 11 post-mission press conference transcript. "The First Lunar Landing, 20th Anniversary," 1989. https://history.nasa.gov/ap11ann/FirstLunarLanding/toc.html.

Armstrong, Neil. Interview by Katy Vine. "Walking on the Moon." TexasMonthly.com, July 2009. https://www.texasmonthly.com/articles/walking-on-the-moon/.

————. Interview by Stephen E. Ambrose and Dr. Douglas Brinkley for NASA Johnson Space Center Oral History Project, September 19, 2001.

Armstrong, Neil, Michael Collins, and Edwin E. Aldrin, Jr. *First on the Moon: A Voyage with Neil Armstrong, Michael Collins, Edwin E. Aldrin, Jr.* With Gene Farmer and Dora Jane Hamblin. Boston: Little, Brown, and Company, 1970.

Bales, Steven G. "From Iowa to the Trench—A Guido's Tale." In *From the Trench of Mission Control to the Craters of The Moon*, 223–260. Self-published, 2012.

BBC. "13 Minutes to the Moon." Season 1, Episode 02. "Kids in Control." Presented by Kevin Fong. Aired May 22, 2019. https://www.bbc.co.uk/programmes/w3csz4dk.

————. "13 Minutes to The Moon." Season 1, Episode 09. "Tranquility Base." Presented by Kevin Fong. Aired July 10, 2019. https://www.bbc.co.uk/programmes/w3csz4ds.

Beattie, Donald A. *Taking Science to the Moon: Lunar Experiments and the Apollo Program.* Baltimore: The Johns Hopkins University Press, 2001.

Bennett, Floyd. "Apollo Experience Report: Mission Planning for Lunar Module Decent and Ascent." Houston: Manned Spacecraft Center, 1972. https://www.hq.nasa.gov/alsj/nasa-tnd-6846pt.1.pdf.

————. Interview by Jennifer Ross-Nazzal for NASA Johnson Space Center Oral History Project, October 22, 2003.

Benson, Charles D., and William B. Faherty. *Moonport: A History of Apollo Launch Facilities and Operations.* Washington, DC: National Aeronautics and Space Administration, 1978.

Bilstein, Roger E. *Stages to Saturn, A Technological History of the Apollo/Saturn Launch Vehicles.* Washington: National Aeronautics and Space Administration, 1980.

Bobst, Kristen. "Astronaut Snoopy's 50-Plus Year History with NASA," July 17, 2019. https://www.pbssocal.org/science/snoopy-has-been-working-with-nasa-since-the-late-1950s/.

Borman, Frank. *Countdown: An Autobiography.* With Robert J. Serling. New York: Silver Arrow, 1988.

————. Interview by Catherine Harwood for NASA Johnson Space Center Oral History Project, April 13, 1999.

Bostick, Jerry C. Interview by Carol Butler for NASA Johnson Space Center Oral History Project, February 23, 2000.

————. *The Kid from Golden: From the Cotton Fields of Mississippi to NASA Mission Control and Beyond, revised edition.* Bloomington: iUniverse, 2017.

———. "Trench Memories." In *From the Trench of Mission Control to the Craters of the Moon*, 143–182. Self-published, 2012.

Brooks, Courtney G., James M. Grimwood, and Lloyd S. Swenson, Jr. *Chariots for Apollo: A History of Manned Lunar Spacecraft*. Washington, DC: National Aeronautics and Space Administration, 1979.

"California Dinner for Apollo 11 Astronauts, part 1." Audio, 11:46. https://archive.org/details/CaliforniaDinnerForApollo11Astronauts/California+Dinner+for+Astronauts_8-13-1969_side+1_.wav.

"California Dinner for Apollo 11 Astronauts, part 2." Audio, 11:38. https://archive.org/details/CaliforniaDinnerForApollo11Astronauts/California+Dinner+for+Astronauts_8-13-1969_side+2_.wav.

Calle, Chris. *Celebrating Apollo 11: The Artwork of Paul Calle*. AeroGraphics, 2009.

Cassutt, Michael. *The Astronaut Maker: How One Mysterious Engineer Ran Human Spaceflight for a Generation*. Chicago: Chicago Review Press, 2018.

CBS News. "The Moon Above, The Earth Below." CBS News. Written and produced by Perry Wolff, hosted by Dan Rather and Charles Kuralt. Aired July 13, 1989. video, 1:34:39.

CBSnews.com. "NASA Legends Remember the Nerve-Wracking Moments Before Apollo 11 Landing," July 15, 2019. https://www.cbsnews.com/news/apollo-11-moon-landing-anniversary-nasa-legends-remember-the-nerve-wracking-moments/.

Cernan, Eugene. *The Last Man on the Moon*. With Don Davis. New York: St. Martin's Press, 1999.

Chaikin, Andrew. *A Man on the Moon: The Voyages of the Apollo Astronauts*. New York: Viking, 1994.

———. *Voices from The Moon: Apollo Astronauts Describe Their Lunar Experiences*. With Victoria Kohl. New York: The Penguin Group, 2009.

Collins, Michael. *Carrying the Fire: An Astronaut's Journeys*. New York: Farrar, Strauss, and Giroux, 1974.

———. Interview by Michelle Kelly for NASA Johnson Space Center Oral History Project. October 8, 1997.

———. *Liftoff: The Story of America's Adventure in Space.* New York: Grove Press, 1988.

Compton, William David. *Where No Man Has Gone Before: A History of Apollo Lunar Exploration Missions.* Washington, DC: National Aeronautics and Space Administration, 1989.

Cooper, Henry S. F., Jr. *Apollo On the Moon.* New York: Dial Press, 1969.

———. *Moon Rocks.* New York: Dial Press, 1970.

Cortright, Edgar M., ed. *Apollo Expeditions to the Moon.* Washington, DC: National Aeronautics and Space Administration, 1975.

Deiterich, Charles F. Interview by Rebecca Wright for NASA Johnson Space Center Oral History Project, February 28, 2006.

Dethloff, Henry C. *Suddenly, Tomorrow Came: The History of the Johnson Space Center.* New York: Dover Publications, Inc., 2012.

Dinn, Mike. Interview by Colin Mackellar for honeysucklecreek.net. Audio, 13:32. https://honeysucklecreek.net/audio/interviews/Mike_Dinn_up_to_A1 1.mp3.

Duke, Charles M. Jr., Interview by Doug Ward for NASA Johnson Space Center Oral History Project, March 12, 1999.

Duke, Charles, and Michael Jones. "Human Performance During a Simulated Mid-Course Navigation Sighting." Master's thesis. Cambridge: Massachusetts Institute of Technology, 1964. https://authors.library.caltech.edu/5456/1/hrst.mit.edu/hrs/apollo/publ ic/archive/1130.pdf.

Duke, Charlie. "Walk on The Moon/Walk with the Son." Christ Our King Anglican Church, January 9, 2019. Video, 1:00:22. https://www.youtube.com/watch?v=Gr-UD6pk6gw&t=2909s.

Duke, Charlie, and Dotty Duke. *MoonWalker: The True Story of An Astronaut Who Found That the Moon Wasn't High Enough to Satisfy His Desire for Success*. Nashville: Thomas Nelson, 1990.

England, Tony. "Human Spaceflight: A History of Competing Objectives." Mardigian Library Lecture Series. Lecture, University of Michigan-Dearborn, February 15, 2019. Video, 1:36:50. https://library.umd.umich.edu/lecture/.

English, Robert A., Kelsey-Seybold Clinic, Richard E. Benson, J. Vernon Bailey, and Charles M. Barnes. "Apollo Experience Report: Protection Against Radiation." Houston: Manned Spacecraft Center, 1973.

Ertel, Ivan D., and Roland W. Newkirk. *The Apollo Spacecraft: A Chronology, Volume IV: January 21, 1966–July 13, 1974*. With Courtney G. Brooks. Washington, D.C.: National Aeronautics and Space Administration, 1978.

Evans, Michelle. *The X-15 Rocket Plane, Flying the First Wings Into Space*. Lincoln: University of Nebraska Press, 2013.

Eyles, Don. *Sunburst and Luminary: An Apollo Memoir*. Boston: Fort Point Press, 2019.

———. "Tales from the Lunar Module Guidance Computer." A paper presented to the 27th annual Guidance and Control Conference of the American Astronautical Society (AAS), in Breckenridge, Colorado, February 6, 2004. https://www.doneyles.com/LM/Tales.html.

Fendell, Edward I. Interview by Kevin M. Rusnak for NASA Johnson Space Center Oral History Project, October 19, 2000.

French, Francis, and Colin Burgess. *In the Shadow of the Moon: A Challenging Journey to Tranquility, 1965–1969*. Lincoln: University of Nebraska Press, 2007.

———. *Into That Silent Sea: Trailblazers of the Space Era, 1961–1965*. Lincoln: University of Nebraska Press, 2007.

Garman, John R. Interview by Kevin M. Rusnak for NASA Johnson Space Center Oral History Project. March 27, 2001.

"Gemini: It Didn't Last Long Enough," *Newsweek*, April 5, 1965. Vol. LXV, No. 14.

"Gen. Samuel C. Phillips: A retrospective on the director of the Apollo program and former commander of the Space and Missile Systems Organization." Space and Missile Systems Center. https://www.losangeles.spaceforce.mil/News/Article-Display/Article/1917223/gen-samuel-c-phillips-a-retrospective-on-the-director-of-the-apollo-program-and/.

Godwin, Robert, ed. *Apollo 9: The NASA Mission Reports*. Ontario: Apogee Books, 1999.

Greene, Jay H. Interview by Sandra Johnson for NASA Johnson Space Center Oral History Project, November 10, 2004.

Griffin, Gerald D. Interview by Doug Ward for NASA Johnson Space Center Oral History Project, March 12, 1999.

Grissom, Gus, and John Young. "Molly Brown was OK from the first time we met her." *Life Magazine*, April 2, 1965. Vol. 58, No. 13.

Guideposts. "Guideposts Classics: When Buzz Aldrin Took Communion on The Moon." Accessed November 9, 2020. https://www.guideposts.org/better-living/life-advice/finding-life-purpose/guideposts-classics-when-buzz-aldrin-took-communion-on-the-moon.

Hacker, Barton C., and James M. Grimwood. *On the Shoulders of Titans: A History of Project Gemini*. Washington: National Aeronautics and Space Administration, 1977.

Haise, Fred W., Jr. Interview by Doug Ward for NASA Johnson Space Center Oral History Project, March 23, 1999.

Hansen, James R. *First Man: The Life of Neil A. Armstrong*. New York: Simon and Schuster, 2005.

Harford, James. *Korolev: How One Man Masterminded the Soviet Drive to Beat America to the Moon*. New York: John Wiley & Sons, Inc., 1997.

Harland, David M. *Exploring the Moon: The Apollo Expeditions*, second edition. Chichester, UK, Springer-Praxis, 2008.

Harris, Gordon L. *Selling Uncle Sam*. Hicksville, NY: Exposition Press, 1976.

"Hasselblad In Space." Accessed July 13, 2020. https://www.hasselblad.com/inspiration/history/hasselblad-in-space/.

Herring, Mack R. *Way Station to Space: A History of the John C. Stennis Space Center*. Washington, DC: National Aeronautics and Space Administration, 1997.

Hoag, David. "Apollo Navigation, Control, and Guidance Systems: A Progress Report." Cambridge: The MIT Instrumentation Laboratory, 1969. https://web.mit.edu/digitalapollo/Documents/Chapter6/hoagprogrepo rt.pdf.

Hoare, Callum. "Moon Landing: Neil Armstrong Reveals What Cameras Missed in Unearthed Interview." Express, October 2, 2019. https://www.express.co.uk/news/science/1185497/moon-landing-neil-armstrong-apollo-11-cameras-missed-fake-apollo-11-spt.

Hollis, Tim. "The Heidi Game in Birmingham." *Birmingham Rewound*. Accessed January 15, 2020. https://www.birminghamrewound.com/features/heidi.htm.

Honeycutt, Jay F. Interview by Rebecca Wright for NASA Johnson Space Center Oral History Project, March 22, 2000.

Houston, Rick, and Milt Heflin. *Go, Flight! The Unsung Heroes of Mission Control, 1965–1992*. Lincoln: University of Nebraska Press, 2015.

Hughes, Francis E. "Frank." Interview by Rebecca Wright for NASA Johnson Space Center Oral History Project, September 10, 2013.

Johnson, Gary W. Interview by Rebecca Wright for NASA Johnson Space Center Oral History Project, May 3, 2010.

Johnson, Stephen B. *The Secret of Apollo: Systems Management in American and European Space Programs*. Baltimore: JHUP, 2006.

Jones, Eric M. "Apollo Lunar Surface Journal." https://www.hq.nasa.gov/alsj/.

Jurek, Richard. *The Ultimate Engineer: The Remarkable Life of NASA's Visionary Leader George M. Low.* Lincoln: University of Nebraska Press, 2019.

Kennedy, John F. "Address at Rice University on The Nation's Space Effort," September 12, 1962. JFK Library Foundation. https://www.jfklibrary.org/learn/about-jfk/historic-speeches/address-at-rice-university-on-the-nations-space-effort.

———. "Address to Joint Session of Congress, May 25, 1961." JFK Library Foundation. https://www.jfklibrary.org/learn/about-jfk/historic-speeches/address-to-joint-session-of-congress-may-25-1961.

———. "Remarks at The Dedication of The Aerospace Medical Health Center, San Antonio, Texas," November 21, 1963. JFK Library Foundation. https://www.jfklibrary.org/archives/other-resources/john-f-kennedy-speeches/san-antonio-tx-19631121.

Kohn, Richard H. *Reflections on Research and Development in the United States Air Force.* Washington, DC: Center for Air Force History, 1993.

Kraft, Christopher C. *Flight: My Life in Mission Control.* New York: Dutton, 2001.

Kranz, Gene. 2018 Vice Adm. Donald D. Engen Flight Jacket Night Lecture. National Air and Space Society, Washington, DC, November 8, 2018. https://airandspace.si.edu/events/flight-jacket-night-eugene-kranz.

———. *Failure Is Not an Option: Mission Control from Mercury to Apollo 13 and Beyond.* New York: Berkley Books, 2000.

Kurson, Robert. *Rocket Men: The Daring Odyssey of Apollo 8 and the Astronauts Who Made Man's First Journey to the Moon.* New York: Random House, 2018.

Lattimer, Dick. *All We Did Was Fly to the Moon.* Gainesville, FL: The Whispering Eagle Press, 1985.

Lewis, Richard S. *The Voyages of Apollo: The Exploration of the Moon.* New York: Quadrangle, The New York Times Book Co., 1974.

Lindsay, Hamish. "Apollo 11, 16–25 July, 1969." Honeysucklecreek.net. https://www.honeysucklecreek.net/msfn_missions/Apollo_11_missio n/index.html.

———. *Tracking Apollo to the Moon*. London: Springer, 2001.

Logsdon, John M., ed. *Exploring the Unknown: Selected Documents in the History of the U.S. Civil Space Program, Volume VII: Human Spaceflight: Projects Mercury, Gemini, and Apollo*. With Roger D. Launius. Washington, DC: National Aeronautics and Space Administration, 2008.

———. *The Decision to Go to The Moon: Project Apollo and the National Interest*. Cambridge: MIT Press, 1970.

Lovell, James A, Jr. Interview by Ron Stone for NASA Johnson Space Center Oral History Project, May 25, 1999.

Lovell, Jim, and Jeffrey Kluger. *Lost Moon: The Perilous Voyage of Apollo 13*. Boston: Houghton Mifflin Company, 1994.

Lunney, Glynn S. *Highways into Space: A first-hand account of the beginnings of the human space program*. Self-published, 2014.

———. Interview by Roy Neal for NASA Johnson Space Center Oral History Project, March 9, 1998.

Mackellar, Colin. "The Apollo 11 Television Broadcast." Honeysucklecreek.net. https://honeysucklecreek.net/Apollo_11/index.html.

———. "How the moonwalk was seen live in Western Australia." Honeysucklecreek.net. https://www.honeysucklecreek.net/Apollo_11/A11_TV_Perth.html.

Manto, Cindy Donze. *Michoud Assembly Facility*. Images of America. Charleston: Arcadia Publishing, 2014.

———. *Stennis Space Center*. Images of America. Charleston: Arcadia Publishing, 2018.

Masursky, Harold, G. W. Colton, and Farouk El-Baz, eds. *Apollo Over the Moon: A View From Orbit*. Washington, DC: National Aeronautics and Space Administration, 1978.

Mattingly, Thomas K., II. Interview by Rebecca Wright for Johnson Space Center Oral History Project, November 6, 2001.

———. Interview by Kevin M. Rusnak for Johnson Space Center Oral History Project, April 22, 2002.

Mindell, David A. *Digital Apollo: Human and Machine in Spaceflight.* Cambridge, MIT Press, 2011.

Moseley, Willie G. *Smoke Jumper, Moon Pilot: The Remarkable Life of Apollo 14 Astronaut Stuart Roosa.* Morley, MO: Acclaim Press, 2011.

Muehlberger, William R. Interview by Carol Butler for JSC Oral History Project, November 9, 1999.

Mueller, George E. Interview by Summer Chick Bergen for Johnson Space Center Oral History Project, January 20, 1999.

Murray, Charles, and Catherine Bly Cox. *Apollo: The Race to the Moon.* New York: Simon and Schuster, 1989.

Nafzger, Richard. "Apollo 7 audio as recorded at Corpus Christi." Honeysucklecreek.net. https://www.honeysucklecreek.net/msfn_missions/Apollo_7_mission /index.html.

NASA. "Apollo 7 Mission Report." Houston: Manned Spacecraft Center, 1968. https://www.hq.nasa.gov/alsj/a410/A07_MissionReport.pdf.

———. "Apollo 9 Mission Report." Washington, DC: National Aeronautics and Space Administration, 1969. https://www.hq.nasa.gov/alsj/a410/A09_MissionReport.pdf.

———. "Apollo 10 Technical Crew Debriefing." Houston: Manned Spacecraft Center, 1969. https://history.nasa.gov/afj/ap10fj/pdf/a10-tech-crew-debrief.pdf.

———. "Apollo 11 Lunar Photography." Greenbelt, MD: NASA Goddard Space Flight Center, 1970. https://ntrs.nasa.gov/api/citations/19720010768/downloads/19720010 768.pdf.

———. "Apollo 11 Preliminary Science Report." Washington, DC: National Aeronautics and Space Administration, 1969.

———. "Apollo 11 Technical Crew Debriefing." Volume 1. Houston: Manned Spacecraft Center, 1969. https://www.hq.nasa.gov/alsj/a11/A11TechCrewDebrfV1_ALSJ.pdf.

———. "Apollo 11 Technical Crew Debriefing." Volume 2. Houston: Manned Spacecraft Center, 1969. https://www.hq.nasa.gov/alsj/a11/A11TechCrewDebrfV2_ALSJ.pdf.

———. "Apollo Configuration Management Manual." Washington, DC: National Aeronautics and Space Administration, 1970. https://ntrs.nasa.gov/citations/19800071127.

———. "Apollo Experience Report—Command Module Uprighting System." Houston: Manned Spacecraft Center, 1973. https://ntrs.nasa.gov/api/citations/19730010171/downloads/19730010171.pdf.

———. "Apollo Program Summary Report." Houston: Johnson Space Center, 1975. https://www.history.nasa.gov/apsr/Apollopt5-2.pdf.

———. "Apollo/Saturn V Launch Complex 39 Mobile Launcher Service Arms Operations and Maintenance." https://ntrs.nasa.gov/api/citations/19700007600/downloads/19700007600.pdf.

———. "Apollo/Saturn Postflight Trajectory," October 6, 1969. https://archive.org/details/nasa_techdoc_19920075301/page/n13/mode/2up.

———. *Apollo Spacecraft News Reference: Command and Service Modules.* Prepared by the Space Division of North American Rockwell Corporation, Downey, CA, in cooperation with NASA's Manned Spacecraft Center. Los Angeles: Periscope Film, LLC, 2011.

———. *Apollo Spacecraft News Reference: Lunar Excursion Module.* Prepared by Public Affairs, Space, at Grumman Aerospace Corporation, Bethpage, New York, in cooperation with NASA's Manned Spacecraft Center. Los Angeles: Periscope Film, LLC, 2011.

———. *Astronautics and Aeronautics 1969.* Washington, DC: National Aeronautics and Space Administration, 1970.

————. *Astronautics and Aeronautics 1970*. Washington, DC: National Aeronautics and Space Administration, 1972.

————. *Astronautics and Aeronautics, Chronology on Science, Technology, and Policy, 1970*. Washington, DC: National Aeronautics and Space Administration, 1972. https://history.nasa.gov/AAchronologies/1970.pdf.

————. "Gemini X Technical Debriefing." Houston: Manned Spacecraft Center, 1966. https://digitalsc.lib.vt.edu/Ms1989-029/Ms1989-029_B05_F4a.

————. "Report of the Apollo 204 Review Board to The Administrator, National Aeronautics and Space Administration." https://www.history.nasa.gov/Apollo204/summary.pdf.

————. *Roundup*, Vol 7, No 14, April 26, 1968. Houston: Manned Spacecraft Center Public Affairs Office, 1968.

————. *Roundup*. Vol 8, No 15, May 16, 1969. Houston: Manned Spacecraft Center Public Affairs Office, 1969.

————. "Saturn V Launch Vehicle Flight Evaluation Report: AS-501 Apollo 4 Mission." Huntsville: George C. Marshall Space Flight Center, 1968. https://archive.org/details/nasa_techdoc_19900066482/page/n85/mode/2up.

————. "The Manned Space Flight Network for Apollo." Greenbelt, Maryland: Goddard Space Flight Center, 1968. https://web.mit.edu/digitalapollo/Documents/Chapter8/apollomsfn.pdf.

————. "The National Aeronautics and Space Act of 1958 (Unamended)," Sec. 102 (a). https://history.nasa.gov/spaceact.html.

————. "The Phillips Report, 1965–1966," NASA Historical Reference Collection. https://www.history.nasa.gov/Apollo204/phillip2.html.

————. "What Made Apollo a Success." Washington, DC: National Aeronautics and Space Administration, 1970. https://ntrs.nasa.gov/api/citations/19720005243/downloads/19720005243.pdf.

NASA FACTS. "Telemetry." Washington, DC: National Aeronautics and Space Administration, 1967.

NASA Public Affairs Office, Kennedy Space Center. *The Kennedy Space Center Story*. Orlando: Graphic House, Inc., 2008.

National Weather Service. "SMG Weather History: Apollo Program." Spaceflight Meteorology Group. https://www.weather.gov/smg/apollo.

Neal, Roy. *Ace in the Hole: The Story of the Minuteman Missile*. Garden City, NY: Doubleday & Company, Inc., 1962.

Newton, Sir Isaac. *A Treatise of the System of the World*. Mineola, NY: Dover Publications, Inc., 1969.

Nixon, Richard. "Remarks to Apollo 11 Astronauts Aboard the *USS Hornet* Following Completion of Their Lunar Mission." The American Presidency Project, 1969. https://www.presidency.ucsb.edu/documents/remarks-apollo-11-astronauts-aboard-the-uss-hornet-following-completion-their-lunar.

O'Hara, Dee. Interview by Rebecca Wright for NASA Johnson Space Center Oral History Project, April 23, 2002.

Orloff, Richard W. *Apollo By the Numbers: A Statistical Reference*. Washington, DC: National Aeronautics and Space Administration, 2000.

Orloff, Richard W., and David M. Harland. *Apollo: The Definitive Sourcebook*. Chichester, UK: Springer-Praxis, 2006.

Phillips, General Sam. Interview by Martin Collins for the National Air and Space Museum. September 28, 1989. Tape 1, Side 1, accessed July 13, 2020, https://airandspace.si.edu/research/projects/oral-histories/TRANSCPT/PHILLIP6.HTM.

———. Interview by Lt Col J.B. Kump for the U.S. Air Force Oral History Project, February 22, 1989. Part 4 of 6. https://commons.erau.edu/phillips-oral-history/.

Phinney, William C. *Science Training History of the Apollo Astronauts.* Washington, DC: National Aeronautics and Space Administration, 2015. https://www.lpi.usra.edu/lunar/strategies/Phinney_NASA-SP-2015-626.pdf.

Qantas. "Onboard the Qantas Flight Which Was Out of This World." July 19, 2019. https://www.qantasnewsroom.com.au/roo-tales/onboard-the-qantas-flight-which-was-out-of-this-world.

Ransford, Gary, Wilbur Wallenhaupt, and Robert Bizzell. "Lunar Landmark Locations—Apollo 8, 10, 11, and 12 Missions." Houston: Manned Spacecraft Center, 1970. https://ntrs.nasa.gov/api/citations/19710002567/downloads/19710002567.pdf.

Reed, H. David. "Looking Back." In *From the Trench of Mission Control to the Craters of the Moon.* 183–200. Self-published, 2012.

Rogers, James. "Apollo 11's Michael Collins Recounts the Crew's Three-Week Quarantine on Their Return from The Moon." Fox News, July 22, 2019. https://www.foxnews.com/science/apollo-11-michael-collins-quarantine-moon.

Rose, Rodney G. Interview by Kevin M. Rusnak for NASA Johnson Space Center Oral History Project, November 8, 1999.

Rush, Martin and Walter Vodges. "Three-Degree-of-Freedom Simulation of Gemini Reentry Guidance." In *The Space Congress Proceedings 2.* https://commons.erau.edu/cgi/viewcontent.cgi?article=3224&context=space-congress-proceedings.

Safire, William. *Before the Fall: An Inside View of the Pre-Watergate White House.* Garden City, New York: Doubleday, 1975.

Schaber, Gerald G. *The U.S. Geological Survey, Branch of Astrogeology—A Chronology of Activities from Conception through the End of Project Apollo (1960–1973).* Reston, VA: US Geological Survey, 2005. https://pubs.usgs.gov/of/2005/1190/of2005-1190.pdf.

Schindler, Kevin, and William Sheehan. *Northern Arizona Space Training.* Images of America. Charleston: Arcadia Publishing, 2017.

Schmitt, Harrison H. "Jack." Interview by Carol Butler for NASA Johnson Space Center Oral History, July 14, 1999.

Schmitt, Joe W. Interview by Michelle T. Buchanan for NASA Johnson Space Center Oral History Project, July 1997.

Schultz, David, Hilary Ray, Eugene Cernan, and Antoine Smith. "Body Positioning and Restrains During Extravehicular Activity." Presented to the Gemini Summary Conference, February 1–2, 1967. https://ntrs.nasa.gov/api/citations/19680005472/downloads/19680005472.pdf.

Scott, David, and Alexei Leonov. *Two Sides of the Moon*. New York: Thomas Dunne Books, 2004.

Scott, David Meerman, and Richard Jurek. *Marketing the Moon: The Selling of the Apollo Lunar Program*. Cambridge: The MIT Press, 2014.

Scouts. "World Scouting Salutes Neil Armstrong," January 1, 2012. https://www.scout.org/node/9739.

Shaffer, Philip C. Interview by Carol Butler for NASA Johnson Space Center Oral History Project, January 25, 2000.

Shayler, David J., and Colin Burgess. *NASA's Scientist-Astronauts*. Chichester, UK: Springer-Praxis, 2007.

Shoemaker, Eugene M. 1963. Interview by Ron Doel for American Institute of Physics Oral History, Session II, June 16, 1987. https://www.aip.org/history-programs/niels-bohr-library/oral-histories/5082-2.

Silver, Leon T. Interview by Carol Butler for NASA Johnson Space Center Oral History Project, May 5, 2002.

Slayton, Donald K., and Michael Cassutt. *Deke! U.S. Manned Space: From Mercury to The Shuttle*. New York: Tom Doherty Associates, 1994.

Spacecraft Films. 2005. "Apollo 7: Shakedown Cruise." Charlotte: Red Pepper Collective, 2005.

Stachurski, Richard. *Below Tranquility Base: An Apollo Memoir*. North Charleston, South Carolina: CreateSpace Independent Publishing Platform, 2013.

Stafford, Thomas P. *We Have Capture: Tom Stafford and the Space Race.* With Michael Cassutt. Washington: Smithsonian Books, 2002.

Stewart, Troy M. Interview by Carol Butler for NASA Jonson Space Center Oral History Project, September 21, 1998.

Styles, Helen. "Boeing Saw Re-entry Spectacle." *Sydney Sun-Herald,* July 27, 1969. Honeysucklecreek.net. https://www.honeysucklecreek.net/images/Apollo_11/Sun-Herald_27-7-69.jpg.

Tindall, Bill. "A lengthy status report on lunar point landing including some remarks about CSM DOI." Tindallgram, August 29, 1969. http://www.collectspace.com/resources/tindallgrams/tindallgrams02.pdf.

————. "Apollo 12 Descent—Final Comments." Tindallgram, November 4, 1969. http://www.collectspace.com/resources/tindallgrams/tindallgrams02.pdf.

————. "DPS low level propellant light." Tindallgram, May 29, 1969. http://www.collectspace.com/resources/tindallgrams/tindallgrams02.pdf.

————. "G Lunar Surface stuff is still incomplete." Tindallgram, March 7, 1969. http://www.collectspace.com/resources/tindallgrams/tindallgrams02.pdf.

————. "How the MSFN and sextant data are used to target DOI and Descent.," Tindallgram, April 16, 1969. http://www.collectspace.com/resources/tindallgrams/tindallgrams02.pdf.

————. "How to land next to a Surveyor: a short novel for do-it-yourselfers." Tindallgram, August 1, 1969. http://tindallgrams.net/69-PA-T-114A.

————. "Some 'Improvements' in Descent Preparation Procedures." Tindallgram, May 15, 1969. http://www.collectspace.com/resources/tindallgrams/tindallgrams01.pdf.

———. "Vent, bent, descent, lament." Tindallgram, August 28, 1969. https://finleyquality.net/wp-content/uploads/2019/07/HSI-39249-Tindall-1969-08-28-Vent-bent-descent-lament.pdf.

Twombly, Matthew. "A Timeline Of 1968: The Year That Shattered America." *Smithsonian Magazine*, January/February 2018. https://www.smithsonianmag.com/history/timeline-seismic-180967503/.

Ward, Jonathan H. *Rocket Ranch: Countdown to a Moon Launch: Preparing Apollo for Its Historic Journey*. Chichester, UK: Springer-Praxis, 2015.

———. *Rocket Ranch: The Nuts and Bolts of the Apollo Moon Program at Kennedy Space Center*. Chichester, UK: Springer-Praxis, 2015.

Wilford, John Noble. *We Reach the Moon*. New York: Bantam Books, 1969.

Wilhelms, Don E. *To A Rocky Moon: A Geologist's History of Lunar Exploration*. Tucson: The University of Arizona Press, 1993.

Wood, Bill. Interview by Colin Mackellar. Honeysucklecreek.net, April 8, 2010. https://honeysucklecreek.net/interviews/index.html.

Woods, David. "Apollo Flight Journal." https://history.nasa.gov/afj/index.html.

———. *How Apollo Flew to the Moon, 2nd edition*. Chichester, UK: Springer-Praxis, 2011.

Young, John W. *Forever Young: A Life of Adventure in Air and Space*. With James R. Hansen. Gainesville: University Press of Florida, 2012.

Lunar Day 17, sketch by Erika Rix.
Courtesy of Erika Rix.

Index

About the Author

With Apollo 16 Command Module Casper at the US Space and Rocket Center, Huntsville, Alabama, October 3, 2021.

Rob Bailey was born and raised in Augusta, Georgia. He graduated from the University of Georgia, where he was a football letterman in 1985. He received a Master of Divinity degree from Reformed Theological Seminary in Jackson, Mississippi, and has been ordained in the Presbyterian Church in America (PCA) since 1996. Rob currently serves as Pastor of Southside Community Church in Corpus Christi, Texas. He and his wife Scotty have been married since 1989 and have four sons: Robert, Alex, Nikolas, and Cameron. He is a lifelong fan of the space program and also enjoys astronomy, hiking, amateur radio, and college football. He can be contacted at rob@apollo16project.org.

VERLORT (Very Long-Range Tracking) radar base from the Mercury era.
Rodd Field Tracking Station, Corpus Christi, Texas.
Photo by author.

Printed in Great Britain
by Amazon

21979603R00229